Studies in Computational Intelligence

Volume 767

Series editor

Janusz Kacprzyk, Polish Academy of Sciences, Warsaw, Poland
e-mail: kacprzyk@ibspan.waw.pl

The series "Studies in Computational Intelligence" (SCI) publishes new developments and advances in the various areas of computational intelligence—quickly and with a high quality. The intent is to cover the theory, applications, and design methods of computational intelligence, as embedded in the fields of engineering, computer science, physics and life sciences, as well as the methodologies behind them. The series contains monographs, lecture notes and edited volumes in computational intelligence spanning the areas of neural networks, connectionist systems, genetic algorithms, evolutionary computation, artificial intelligence, cellular automata, self-organizing systems, soft computing, fuzzy systems, and hybrid intelligent systems. Of particular value to both the contributors and the readership are the short publication timeframe and the world-wide distribution, which enable both wide and rapid dissemination of research output.

More information about this series at http://www.springer.com/series/7092

Jamshaid Ashraf · Omar K. Hussain
Farookh Khadeer Hussain
Elizabeth J. Chang

Measuring and Analysing the Use of Ontologies

A Semantic Framework for Measuring Ontology Usage

 Springer

Jamshaid Ashraf
Digital Processing System (DPS)
Salmiya
Kuwait

Omar K. Hussain
School of Business
University of New South Wales
Canberra, ACT
Australia

Farookh Khadeer Hussain
Centre for Artificial Intelligence, Faculty of
  Engineering and Information Technology,
  School of Software
University of Technology Sydney
Sydney, NSW
Australia

Elizabeth J. Chang
School of Business
University of New South Wales
Canberra, ACT
Australia

ISSN 1860-949X   ISSN 1860-9503   (electronic)
Studies in Computational Intelligence
ISBN 978-3-030-09296-2   ISBN 978-3-319-75681-3   (eBook)
https://doi.org/10.1007/978-3-319-75681-3

Printed on acid-free paper

This Springer imprint is published by Springer Nature
The registered company is Springer International Publishing AG
The registered company address is: Gewerbestrasse 11, 6330 Cham, Switzerland

I would like to thank my late parents for their sacrifices and support. I am indebted to my elder brother for helping me to get this far. I am grateful for the continuous support and encouragement of my wife, daughters and son. Above all, thanks to the Almighty Allah for His uncountable blessings.

—Jamshaid

I am indebted to my parents for the sacrifices they made in helping me to reach a stage where I am at present. I also appreciate and am grateful for the support and constant encouragement of my wife and my lovely daughters in all my endeavors. Above all, I thank the Almighty Allah for everything he has given me in my life.

—Omar

I would thank my dearest parents, dearest wife, and my dearest sons for all their support and help.

—Farookh

I would like to thank my late mother for her inspirational up-bringing and support.

—Elizabeth

Foreword

Since the Gruber's definition in 1990s on ontology for use in computer science, the theory and practice on ontology in information sciences has grown in a remarkable way. Nowadays, the set of ontology definitions and uses have become a cross-cutting feature of multidisciplinary application domains by embracing fields of computer science, software engineering, knowledge discovery, inference reasoning systems and decision support systems, semantic social networks, context-aware systems, enterprise systems, to name a few. Indeed, ontology provides powerful tools to formally describing concepts and relationships, modelling knowledge domain and application behaviour, understanding the context and relationships among various contexts, etc., based on the use of domain ontology leading to structural frameworks for organizing information. Obviously ontology theory has been fuelled by the needs of semantic technology and has become an inseparable part of semantic world (also referred to as 'Ontological Semantic Technology' or 'Ontological Semantics')!

The book *Measuring and Analysing the Use of Ontologies* is an excellent work putting together all the pieces of ontology theory and practice. It covers in a comprehensive way the concepts, definitions and uses of ontology as well as measuring their usefulness in the development of computer systems. Thus, the book covers in great detail the lifecycle of ontologies, from their conception and formal definition to their evaluation and validation.

The authors of the book bring their extensive research experience and expertise in the field of ontology, knowledge-based systems and semantic technologies. The readers of this book can benefit by the full coverage of the latest advances on the qualitative and quantitative analysis frameworks for measuring ontology usage as well as their application to many use cases arising in real-life application and knowledge domain scenarios. Lessons learned, future research directions and challenges are also presented, which will be particularly of interest to young researchers and developers in the ontology field.

The authors have adopted a good writing style yet a detailed technical approach, making the book easy to understand and follow.

I congratulate the authors for their book and wish the readers to enjoy the book!

Barcelona, Spain   Fatos Xhafa, Ph.D.
November 2017   Professor Titular d'Universitat
Departament de Llenguatges i Sistemes Informàtics
Universitat Politècnica de Catalunya

Preface

Measurement is the first step that leads to control and eventually to improvement. If you cant measure something, you cant understand it. If you cant understand it, you cant control it. If you cant control it, you cant improve it.

—H. James Harrington

The Semantic Web envisions a Web where information is accessible and processable by computers as well as humans. Ontologies are the cornerstones for realizing this vision of the Semantic Web by capturing domain knowledge by defining the terms and the relationship between these terms to provide a formal representation of the domain with machine-understandable semantics. Ontologies are used for semantic annotation, data interoperability and knowledge assimilation and dissemination. In the literature, different approaches have been proposed to build and evolve ontologies, but in addition to these, one more important concept needs to be considered in the ontology lifecycle, that is, its usage. Measuring the 'usage' of ontologies will help us to effectively and efficiently make use of semantically annotated structured data published on the Web (formalized knowledge published on the Web), improve the state of ontology adoption and reusability, provide a usage-based feedback loop to the ontology maintenance process for a pragmatic conceptual model update, and source information accurately and automatically which can then be utilized in the other different areas of the ontology lifecycle. In today's age of Big Data, having such information that will help us to make sense out of Big RDF data arising from the increase of Semantic Web's popularity is important. Ontology Usage Analysis (OUA) is the area which evaluates measures and analyses the use of ontologies on the Web. However, in spite of its importance, no formal approach is present in the literature which focuses on measuring the use of ontologies on the Web. This is in contrast to the approaches proposed in the literature on the other concepts of the ontology lifecycle, such as ontology development, ontology evaluation and ontology evolution. So, this book is an effort to address this gap in the literature by justifying and proposing a comprehensive framework to assess, analyse and represent the use of ontologies on the Web.

In order to achieve the aforesaid objective, this book is divided into 10 Chapters. In Chap. 1, the growth of Semantic Web, ontologies in different domains and ontology lifecycle are discussed before justifying the motivation for OUA. Different use case scenarios from the viewpoint of a user in different roles are taken to explain the importance of OUA in the ontology lifecycle. Chapter 2 discusses the ontology-focused work from the literature in detail. The discussion in this chapter is divided in to the two viewpoints of *Usage and Ontology*, to identify the existing gaps in the literature related to OUA. Finally, this chapter concludes by highlighting what is missing with respect to OUA and positioning it in the ontology lifecycle. OUA is analysed and discussed with relevant areas such as ontology evaluation and ontology evolution to discuss the function and scope of OUA and highlight the subtle differences it has between other areas of ontology engineering. In Chap. 3, the key concepts followed by a comprehensive operational definition for OUA are defined, taking into account the different objects that are of interests during the analyses of domain ontologies. Following the definition, an approach for OUA namely the <u>O</u>ntology <u>US</u>age <u>A</u>nalysis <u>F</u>ramework (OUSAF) is presented, and the different phases in that approach namely identification, investigation, representation and utilization for carrying out usage analysis are discussed. Comprehensive details on each phase of the framework along with their objectives and functional requirements are explained before concluding the chapter.

In Chaps. 4–9, the process of OUA by the proposed OUSAF is discussed. In Chap. 4, a methodological approach for the identification phase of the OUSAF is presented. Various metrics that are important for the identification process to be carried out are defined before proposing the Ontology Usage Network Analysis Framework (OUN-AF) for the implementation of the identification phase. The OUN-AF comprises three phases: the input phase, computation phase and analysis phase. The input phase is responsible for collecting the data for ontology identification. In the computation phase, the ontology usage network is constructed to provide the computational architecture for observing the relationships that ontologies have with data sources. In the analysis phase, the defined metrics are computed by using a case study to demonstrate the observed relationship between ontologies and the data sources using those ontologies. Chapters 5 and 6 focus on the investigation phase of the OUSAF which is at two levels, first at an empirical level in which ontologies are empirically analysed and second at a quantitative level, in which ontology usage is quantitatively analysed and measured. Chapter 5 discusses the EMPirical Analysis Framework (EMP-AF) to empirically analyse the use of domain ontologies on the Web. The EMP-AF comprises two phases: the data collection phase and the aspect analysis phase. In the data collection phase, data is collected from the Web and in the aspect analysis phase, usage is analysed from different aspects using the proposed metrics. Chapter 6 discusses the QUAntitative Analysis Framework (QUA-AF) for the quantitative level investigation phase of the OUSAF. QUA-AF comprises three phases: the data collection phase, the computation phase and the application phase. In the data collection phase, the data is collected by crawling the Web. In the computation phase, for each dimension, a different set of metrics is defined and their results are then consolidated to obtain a unified rank of the usage. The frameworks in each chapter are explained in detail along with their key concepts, the different metrics used in each of them, different use cases according to the type of user being considered and then demonstrating the output that is achieved from each of them related to the OUA. Chapter 7 focusses on the representation phase of the OUSAF wherein formalization of OUA domain knowledge for the representation phase of the outputs of the different phases of the OUSAF is represented. A conceptual model known as the *UOntology* is built to represent domain knowledge. The conceptual model is then formalized on UML which is a standard modelling language, and is implemented using a formal ontology language to generate the ontology artefact for population and further utilization. In Chap. 8, the focus is on the utilization phase of OUSAF to demonstrate the benefits of OUA obtained from the proposed framework. Different use cases according to the different types of users are identified and by using the proposed frameworks from Chaps. 4–7, the implementation of those use case scenarios is shown to demonstrate the usefulness of the results OUA obtained from the OUSAF. In Chap. 9, the developed conceptual model (UOntology) to represent the domain knowledge while OUA is evaluated to demonstrate the validity of the results that it represents from the OUSAF.

Chapter 10 concludes the discussions in this book by discussing briefly the future research directions.

Salmiya, Kuwait   Jamshaid Ashraf
Canberra, Australia   Omar K. Hussain
Sydney, Australia   Farookh Khadeer Hussain
Canberra, Australia   Elizabeth J. Chang

Acknowledgements

The first author acknowledges the support received from Mohammad Amjad Ch, CEO of DPS Kuwait during the research and book compilation work. All authors acknowledge Ms Sue Felix for proofreading the chapters of the book. They also express their deepest gratitude to Ms Monireh Alsadat Mirtalaie for her help in different tasks while compiling the book. Finally, their appreciation also goes to the Springer editorial staff for their help during the typesetting process.

Contents

Acronyms

API	Application Programming Interface
B2B	Business to Business
B2C	Business to Consumer
CUT	Concept Usage Template
DC	Dublin Core
EHR	Electronic Health Record
EMP-AF	Empirical Analysis Framework
FOAF	Friend Of A Friend
GR	GoodRelations
GRDS	GoodRelations Dataset
GRO	GoodRelations Ontology
HCLS	Health care and Life Science
HTML	HyperText Markup Language
HTTP	Hypertext Transfer Protocol
ICD	International Classification of Diseases
LOC	Linked Open Commerce
LOD	Linked Open Data
OAF	Ontology Application Framework
OBO	Open Biological Ontology
ODLC	Ontology Development Life Cycle
OE	Ontology Evaluation
OMV	Ontology Metadata Vocabulary
OUA	Ontology Usage Analysis
OUD	Ontology Usage Distribution
OUN	Ontology Usage Network
OUN-AF	Ontology Usage Network Analysis Framework
OUSAF	Ontology Usage Analysis Framework
OWL	Web Ontology Language
OWL-DL	Web Ontology Language – Description Logic
PLD	Pay Level Domain

QUA-AF	Quantitative Analysis Framework
RDF	Resource Description Framework
RDFa	Resource Description Framework Attribute
RDFS	Resource Description Framework Schema (RDF Schema)
SDLC	Software Development Life Cycle
SKOS	Simple Knowledge Organization System
SLG	Schema Link Graph
SNA	Social Network Analysis
SPARQL	SPARQL Protocol and RDF Query Language
UML	Unified Modelling Language
URI	Uniform Resource Identifier
URL	Uniform Resource Locator
W3C	World Wide Web Consortium
WWW	World Wide Web
XML	Extensible Mark-up Language

List of Figures

List of Tables

Chapter 1
Motivation

1.1 Introduction

[1]Since its inception, the internet has transformed the way we communicate, interact and do business across the globe. Described and dubbed as *information highway*, the internet has provided an unprecedented seamless infrastructure to assimilate and dissimilate information with an ease and speed never before witnessed by mankind. Today, as a result of this, 51.7% of the world's population is using the *internet*.[2] Various recent initiatives such as promotion and adoptions of internet in remote regions and under privileged areas has increased the actual internet usage than the reported figures. Capitalizing on such intrinsic properties of the internet as simplicity, ubiquity and scalability, Tim Berners-Lee introduced the World Wide Web in 1989 [25] as a platform for publishing and consuming information on a universal scale. The World Wide Web (also known as the WWW or *Web*), which without a doubt is one of the most significant computational phenomena to date, has revolutionized information sharing by providing a decentralized information platform which has enabled and empowered users to be more interactive and participative, turning each user of the Web into a potential publisher (Fig. 1.1). Being able to publish information which is accessible to anyone in the world with access to the Web for a low cost has resulted

---

[1]Parts of this chapter have been republished with permission from:

1. Knowledge-Based Systems Volume 80, Jamshaid Ashraf, Elizabeth Chang, Omar Khadeer Hussain, Farookh Khadeer Hussain, Ontology usage analysis in the ontology lifecycle: A state-of-the-art review, pp. 34–47, 2017.

2. John Wiley & Sons from Making sense from Big RDF Data: OUSAF for measuring ontology usage, Jamshaid Ashraf, Omar Khadeer Hussain, Farookh Khadeer Hussain, Volume 45, Issue 8, Copyright ©2014 John Wiley & Sons, Ltd; permission conveyed through Copyright clearance centre.

3. Jamshaid Ashraf, Omar Khadeer Hussain Farookh Khadeer Hussain, A Framework for Measuring Ontology Usage on the Web, The Computer Journal, 2013, Volume 56, Issue 9, pp. 1083–1101, by permission of OUP.

[2]http://www.internetworldstats.com/stats.htm; retr. 16/10/2017.

© Springer International Publishing AG 2018
J. Ashraf et al., *Measuring and Analysing the Use of Ontologies*, Studies in Computational Intelligence 767, https://doi.org/10.1007/978-3-319-75681-3_1

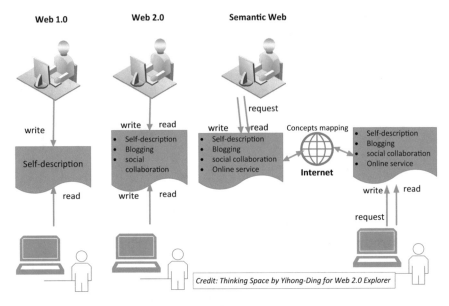

Fig. 1.1   A simple picture of Web Evolution [82]

in the proliferation of approximately 4.62 billion indexed web documents[3] containing information on a variety of topics, and creating a huge amount of diversified
information commonly referred as Big Data.

As a consequence, we are witnessing a persistent rise in user-generated content
that is padded with **metadata** to provide additional (syntactical and structural) information about the content, such as content ownership, provenance detail, content
categorization and labelling. This stage of the evolution of the Web is termed Web
2.0 in the literature [213] and is described *as a concept that takes the internet as a
platform for information sharing, interoperability, user-centred design, and collaboration on the Web* (Fig. 1.1). Recent advancement have taken this notion further with
Web 3.0 where collecting intelligence is desired from the information on the Web
and assist user in their queries. With the ability to interact and participate in content
generation, Web 2.0 has provided the necessary techniques [279] and approaches
(such as Web APIs, mash-ups, blogging software, tagging) to link documents with
users (whether publisher or consumer) by adding meta-information to user-generated
content. One of the major contributions of this evolution is the publication of metadata (describing the content user generate and linking it to them) which, in fact,
signalled the early emergence of structured data on the Web, paving the way for the
next stages of evolution.

While having such meta-information is useful, its full potential can only be realized if we are able to retrieve the required information from the Web to consume
for our individual or collective needs. While it sounds possible, in reality, it is a

---

[3]http://www.worldwidewebsize.com retr. 21/10/2017.

daunting challenge to retrieve accurate information when machines have no cue for understanding the content and structure of web documents. Search engines such as Google, Bing and Yahoo! have worked hard to process and sift through unstructured web documents to classify and index them. This pre-processing of documents, although helping search engines to find and return prioritized lists of query-relevant documents [147], falls short in providing answers to specific queries, which is what is needed; *give me what i want when i want it.* In preprocessing, extensive engineering and algorithmic effort [62, 215] has been exerted to understand the information and provide relevant and useful information to users, but this useful information has been limited to returning prioritized lists of relevant documents because the information represented in web documents does not currently contain the metadata required for machines to understand what the content means.

After realizing the potential of having structured data available on the Web [10], there was a push toward developing more sophisticated approaches to accessing information across the Web with improved accuracy to address these limitations. Thus, search engines applied information processing techniques to go beyond keyword-based search and provide support for more complex and comprehensive queries to allow users to access more precise information. However, the quest to provide answers to complex queries, such as 'finding the doctors in a city specializing in mental health', highlighted the need for a more granular and structured representation of information at the data level. The representation of information at the data level on the Web meant that everyone should be able to access, process and interpret the information consistently and coherently.

To address these challenges and take the Web closer to its original envisaged design[4] Tim Berners-Lee and colleagues [255] proposed the Semantic Web vision in 1998, which is described as:

> I have a dream for the Web [in which computers] become capable of analysing all the data on the Web the content, links, and transactions between people and computers. A 'Semantic Web', which should make this possible, has yet to emerge, but when it does, the day-to-day mechanisms of trade, bureaucracy and our daily lives will be handled by machines talking to machines. The 'intelligent agents' people have touted for ages will finally materialize.

This vision of the Semantic Web (as shown in Fig. 1.1) seeks to transform the present *Web-of-Documents* into a *Web-of-Data* where the Web forms a global space for seamless knowledge integration. This global space provides the mechanism to start describing and then linking tangible and non-tangible entities such as *people, software modules, projects, concepts, documents,* etc., on the Web. This is being achieved by Semantic Web, its core technology stack and Linked Data principles, as described in the next section.

---

[4]http://www.w3.org/History/1989/proposal.html; retr. 10/10/2017.

1.2   Semantic Web

The Semantic Web (also dubbed Web 3.0, the Linked Data Web, and the Web of Data) represents the on-going major evolution of the Web in the form of transforming *data* into *meaning*. This transformation enables data across multiple sources to be linked and processed by computers for further complex cognitive tasks after understanding them. These sophisticated tasks require a knowledge processing capability to realize different applications that come under the rubric of searching, information interoperability, knowledge integration and information retrieval. In order to embrace this major evolution (of the Web) to realize the Semantic Web vision [20], the Semantic Web community has taken steps to standardize the underlying foundational components to make them conform to the original Web architecture [21]. The guiding principles considered when standardizing information representation at a *syntactic* and *semantic* level are as follows:

- Resources are identified using the Unique Resource Identifier (URI) to make them accessible over the Web.
- Resources are described using standard formats to make their access and reference consistent across different consuming applications.
- Resources are represented using a standard data model which is flexible and compatible with Web architecture.
- Resources are semantically described to allow the aggregation and combination of data drawn from a range of resources.

The Resource Description Framework (RDF) [164] is the W3C standard for the representation of data and knowledge on the Web and forms a foundational data model of the Semantic Web. This is shown in Fig. 1.2, and is known as the Semantic Web layer cake, that has various layers responsible for different roles utilizing different technological components. On a high level, RDF provides the means to connect resources (things, data, documents, abstract idea, etc.) in a structured and meaningful way. RDF for the resources creates statements in the form of RDF triples. A RDF triple statement consists an expression in the form of subject-predicate-object about the resource. Technically, RDFS (Resource Description Framework Schema) is the most basic schema language which provides declarative schemata whose semantics are defined within RDF Schema [46]. RDFS extends RDF vocabulary to allow the description of taxonomies of classes and properties to develop lightweight vocabularies. While RDF provides a standard structured data model and RDFS declarative schema, OWL (Web Ontology Language) [74] provides a highly declarative expressive language to formally conceptualize the knowledge of a given domain. OWL by extending RDF and RDFS, aims to empower the Semantic Web by giving it the expressive and reasoning power of description logic. In order to query information that is semantically described and structured using an RDF data model, W3C provides SPARQL as a standard query language for RDF data. It contains the SPARQL protocol and RDF query language to allow users and applications to write queries and to consume the results of queries across distributed sources of information (knowledge bases).

Fig. 1.2 The Semantic Web Stack (SW layer cake)

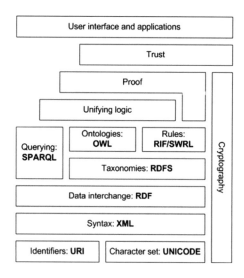

While these technological components provide the standards to implement the Semantic Web vision, they do not provide any guidelines to promote the grass-roots adoption of these standards. To address this issue and accelerate adoption, Tim Berners-Lee, along with the Semantic Web community, introduced **Linked Data** [30] principles to facilitate the publishing and interlinking processes involved in generating semantically rich structured data over the Web. The four Linked Data principles [22] are as follows:

1. Use URIs as the names for things;
2. Use HTTP URI so that names can be dereferenced, i.e. looked up and referred to;
3. Upon look-up, return useful information;
4. Include links by using URIs which link to other related remote documents.

To reap the potential benefits offered by the Semantic Web, many domain-specific industries and their major players, researchers, practitioners and governments have begun to adopt Linked Data principles to disseminate information in a machine-interpretable way. Notable examples are: governments[5] entities [77] such as UK,[6] USA,[7] Australia[8]; corporations such as the BBC [170], New York Times [228], Thompson Reuters [169], Freebase [36], Volkswagen,[9] BestBuy [44]; community-driven Linked Open Data (LOD2) project[10] and DBpedia [11]; biomedical and health-

---

[5]To access the updated and extended list of countries participating in the Open Data initiative, visit http://logd.tw.rpi.edu/ and to obtain the initial analysis visit https://logd.tw.rpi.edu/page/international_dataset_catalog_search; retr. 23/10/2017.

[6]http://data.gov.uk/; retr. 17/08/2017.

[7]http://www.data.gov/; retr. 02/10/2017.

[8]https://data.gov.au; retr. 12/10/2017.

[9]http://www.w3.org/2001/sw/sweo/public/UseCases/Volkswagen/; retr. 21/10/2017.

[10]http://lod2.eu/WikiArticle/Project.html; retr; 21/10/2017.

care data sets such as DrugBank,[11] UniPort,[12] LinkedCT,[13] PubMed[14]; and several other datasets.[15]

However, for machines to interpret information in a common way, distributed ontologies are used to provide machine-processable meta-information which enables automatic information sourcing, retrieval and interlinking. In the next section, we discuss ontologies and their role in Semantic Web in detail.

1.3 Ontologies

Ontologies are the main component of the Semantic Web vision as they provide the semantics for the RDF data; that is, they transform data into meaning. In the literature, ontologies are defined [122] as a *formal specification of conceptualization*. Ontologies are viewed as a shared and common understanding of the domain that can be communicated between people and heterogeneous distributed application systems, as shown in Fig. 1.3 (depiction appeared in [117]). Thus, they specify a machine readable vocabulary in computer systems, which is then used to infer and integrate knowledge, based on the semantics they describe.

Ontologies, which are comprised of concepts, relationships, individuals, and axioms, are constructed to formally conceptualize consensual (shared) knowledge about a particular domain. These components of ontologies are identified by Uniform Resource Identifiers (URI) [23] to offer a global naming scheme. Data publishers use these URIs to describe information in order to promote consistent and coherent semantic interoperability between users, systems and applications. To reap the benefits of the Semantic Web, several domain ontologies have been developed to describe the information pertaining to different domains such as Healthcare and Life Science (HCLS) [69], governments,[16] social spaces [43, 51], libraries [118], entertainment [223], financial services [110] and eCommerce [142].

Numerous use case scenarios have been published in the literature during the last decade in which ontologies have been used in research fields as diverse as Information Retrieval and Extraction, Information Interoperability, Database Design, Entity Disambiguation, Knowledge Representation and Management and Information Modelling [126, 191]. While the early emphasis of ontology exploration was on monolithic systems, a conscious effort by the Semantic Web research community

---

[11] http://www.drugbank.ca/.

[12] http://www.uniport.org.

[13] http://www.linkedct.org.

[14] http://www.ncbi.nlm.nih.gov/entrez/query.fcgi.

[15] Two of the resources which maintain statistics on the different datasets published by following Linked Data principles are: http://stats.lod2.eu/rdfdocs (retr, 26/10/2017) and http://datahub.io (retr. 19/10/2017).

[16] http://oegov.us/ and https://dvcs.w3.org/hg/gld/raw-file/default/dcat/index.html; retr. 12/10/2017.

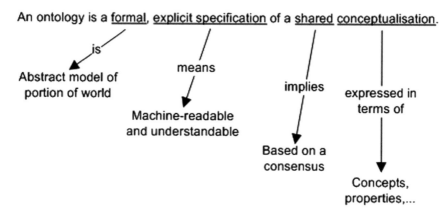

Fig. 1.3 Detailed description of ontology definition [247]

since 2002 has brought the Web into focus for ontology use. This focus on ontologies on the Web has opened up new research fields such as Information Management (particularly in the context of libraries), ontology-enhanced search, ontology driven web navigation, and e-commerce [142].

The research effort in using ontologies in the above-mentioned research areas has helped to develop practical solutions for many application areas. In the subsections that follow, we discuss a few of the application areas and the role of ontologies in them.

1.3.1 Ontologies in Healthcare

One of the first application areas in which ontologies were used is the field of Healthcare and Life-Science (HCLS). Several projects in healthcare have been initiated at government level [171] to provide better health care services by enabling information interoperability across geographical boundaries. Health information interoperability is achieved through the implementation of eHealth initiatives which make use of ICT to facilitate health and healthcare data and transcend socio-political boundaries for borderless healthcare delivery.

Ontologies have been the core component of eHealth initiatives to provide information interoperability on a syntactic as well as a semantic level. A report on Semantic Interoperability for Better Health and Safer Healthcare published by the European Commission [245] recommends the cross-border interoperability of electronic health record systems across Europe with the aim of improving patient health care management. Common examples of healthcare information that is exchanged between diverse systems include patient identification, access to the demographic data of patients, patient medical history, medication records, information stored in Electronic Health Records (EHR), and clinical data (lab tests) [227], etc.

Alan Rector, who is considered an authority in the field of ontologies and health-care, and biomedicine, describes the role of ontologies in healthcare as [224]:

> Ontologies are about the things being represented patients, their diseases. They are about what is always true, whether or not it is known to the clinician. For example, all patients have a body temperature (possibly ambient if they are dead); however, the body temperature may not be known or recorded. It makes no sense to talk about a patient with a "missing" body temperature.

> Data structures are about the artefacts in which information is recorded. Not every data structure about a patient needs to include a field for body temperature, and even if it does, that field may be missing for any given patient. It makes perfect sense to speak about a patient record with missing data for body temperature.

Ontologies, as observed by Alan Rector, are the critical solution component for addressing semantic interoperability in health care systems and facilitating the seam-less integration of the diverse applications and systems that constitute today's health-care technology landscape. As shown in Fig. 1.4, different actors and entities such as health care providers, patients, hospitals, pharmacies, authorities and citizens in general can, using ontologies, interact and converse with each other through the transmission and use of meanings of data and information across different health care services. The ability to associate explicit semantics with data and information not only enables local interoperability, such as within hospitals networks, but also across hospital networks in different countries [245]. Some examples of the types of data exchange that benefit from the semantic interoperability offered by ontologies are:

- Individual patient data can be exchanged across systems and jurisdictions. Elec-tronic health records (EHR), which encompass a systematic collection of electronic health information about an individual patient, are currently at disparate locations at various doctor clinics, hospitals, imaging centres, specialists, and allied health practices. Ontologies help us to describe a patient and his/her health related infor-mation in a format that can be automatically retrieved, processed and seamlessly consumed by a range of applications and systems.
- Collecting health related information from distributed systems and assimilating them to obtain aggregated population data provides the insights required for tack-ling challenging health related issues such as epidemics and the spread of danger-ous viruses. It also assists policy makers and researchers to prepare their future research agendas. Ontologies assist in the collection of data from the diverse sys-tems described using vocabularies and medical terminologies, and interlinking them with relevant pieces of information based on their semantic similarity as well as their context.
- Health care related information described using ontologies provide a rich corpus of health related information to enable the development of knowledge driven health applications and promote collaborative research to tackle major problems such as cancer.

The ability of ontologies to formalize the conceptual component of domain knowledge makes them the perfect solution for the problems mentioned above.

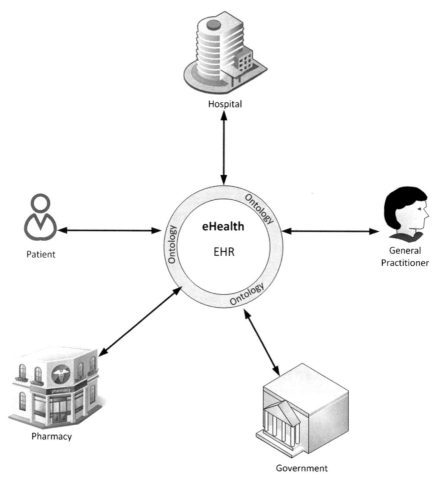

Fig. 1.4 Role of ontologies in eHealth

Government initiatives such as SEMIC - Semantic Interoperability Community,[17] European SemanticHealthNet program[18] and the mapping of several health care and life science related standards such as HL7, ICD with ontologies and Semantic Web technologies [27, 76, 264] demonstrate the value of adopting ontologies in the health care domain.

---

[17]https://joinup.ec.europa.eu/collection/semantic-interoperability-community-semic; (retr. 12/10/2017).

[18]http://www.semantichealthnet.eu/; (retr. 02/11/2017).

1.3.2   Ontologies in E-Commerce

eCommerce (or e-commerce), which describes business that is conducted over the internet using a variety of applications and communication solutions, has been always at the forefront of adopting the latest technologies. Similarly, ontologies, once they had come of age, were adopted in the eCommerce domain. eCommerce essentially requires extensive communication and the exchange of documents between diverse participants and systems and is faced with the lasting issue of heterogeneity. Ontologies, which are largely popular due to their promise to provide a shared and common understanding of a domain that can be communicated between people and application systems, are being used in eCommerce to address the issue of heterogeneity. There are several facets of eCommerce in which ontologies play an important role in Business to Business (B2B) and Business to Consumer (B2C) processes and interaction.

We briefly discuss the role ontologies can play in the above-mentioned facets of eCommerce.

1.3.2.1   Business to Business Transactions

Business to Business (B2B) refers to business that is conducted between companies, rather than between a company and individual consumers. Figure 1.5 shows the exchange of information that is often required between two businesses of a supply chain, namely a manufacturer and distributor.

Manufacturers need to communicate about their products, and product descriptions are the building blocks of an electronic catalogue. The exchange of this information is essential for conducting its business, but it is necessary to deal with the issue of heterogeneity in product descriptions and product catalogues. Ontologies are being used to define formal semantics for information, especially for products, thus allowing the efficient and effective exchange of information across different entities. Ontologies have been used to develop e-Catalogs that are structured, understandable

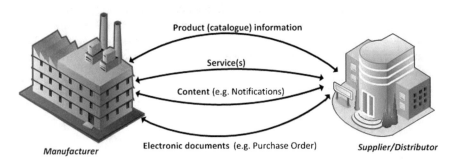

Fig. 1.5 B2B communication model

and processable by computers [176]. The operational implementation of such systems has been applied to government procurement services in which they act as the product ontology knowledge. This knowledge base is then used to design and develop product databases that can be searched and discovered by various applications.

One recent project in this domain, namely the, The Product Type Ontology[19] publishes products and their description data on the Web using Semantic Web technologies in an attempt to enable automatic access to, and processing of, products and their descriptions by other computer programs. Such use of ontologies in Business-2-Business eCommerce is helping to address the bottlenecks highlighted by Fensel et al. [95] as "...we need intelligent solutions for mechanizing the process of structuring, standardizing, aligning, and personalizing data." Ontologies are providing the mechanism required to enable flexible and open eCommerce that is capable of addressing heterogeneity issues by enabling distributed and diverse systems to communicate with and understand products, product descriptions and product catalogue-related information.

1.3.2.2 Business to Consumer Transactions

In contrast to the Business to Business (B2B) model, transactions in Business to Consumer (B2C) are conducted directly between a company and a consumer. Consumers in this model are the end-users of products or services offered by the business. A general B2C model is presented in Fig. 1.6.

In a standard interaction, the consumer - who in this case is the end user - interacts with the eCommerce web site to search for a product or service for possible purchase. An ideal user experience in this interaction can be envisaged only if the information about the products or services can be described and interpreted by the user without ambiguity. This entails the need for specificity to be provisioned in eCommerce systems to enable users (both human and machine) to identify products and products related information unambiguously and autonomously. Ontologies are the ideal technological components for describing information on the Web that is understandable by both humans and machines.

The use of ontologies has allowed a shared understanding to be established that allows the automated processing of information by software agents. The GoodRelations Ontology (GRO) [142], developed specifically for Web-based eCommerce, is an example of an ontology that allows businesses to describe their product offerings, entities and descriptions. The resulting semantically annotated structured data is then accessible for use in Semantic Web applications and inclusion in search engine indexes. GRO provides the concepts to describe products, services, offerings, prices, legal entities, terms and conditions, payments, delivery, and product descriptions. Using GRO or other open ontologies facilitates access to structured semantic data to enable the automatic processing of information, empowering users to access precise and specific information on the Web as a result.

---

[19]http://www.productontology.org/; retr: 12/10/2017.

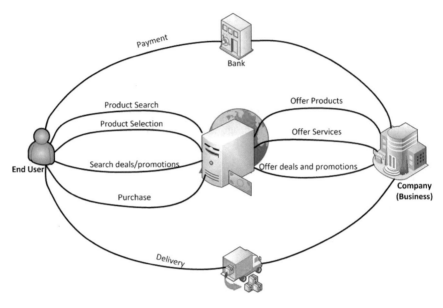

Fig. 1.6 B2C model

1.3.3 Ontologies in Software Engineering

Like other application domains, ontologies are being extensively used in software engineering applications to take advantage of the benefits they offer. In the literature [288], ontologies have been developed to formalise the domain knowledge of the discipline of software engineering as shown in Fig. 1.7. In the domain of software engineering, there are three specific areas in which ontologies can be used to describe information and conceptualize domain knowledge. These are: *domain knowledge of software engineering* [239] which covers a whole set of software engineering concepts; *software development methodology* and its approaches, which might vary across organizations and projects; and the domain knowledge of the particular *application area* for which a software is being developed (e.g. banking, supply chain or medicine).

Semantic descriptions of the software engineering domain, software development methodologies and software artefacts (developed source code) are being used to develop integrated multi-site (distributed) software development platforms [134, 195, 288].

These applications show how ontologies are being used in a wide range of real world applications. As with any information system or product, ontologies as an end product go through different construction stages before they can be used, and this is discussed in the next section.

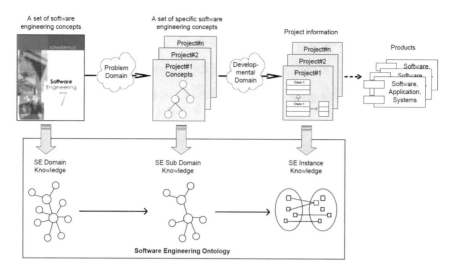

Fig. 1.7 Different areas of software engineering in which ontologies are applicable [288]

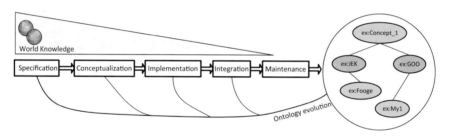

Fig. 1.8 Ontology development and deployment lifecycle process

1.4 How Are Ontologies Developed?

Ontologies have a predictable lifecycle consisting of a number of development and deployment stages. This is termed the Ontology Development and Deployment Lifecycle process (ODDLC) and is similar to the Software Development Lifecycle (SDLC) of software projects [163]. Each stage of the ODDLC progresses the development of the ontology from the conceptual phase to the realization phase, as shown in Fig. 1.8. Before we describe each phase of ODDLC, we define the different users who are the stakeholders in the different stages this process.

- *Ontology developers:* this group of users are involved in the construction (or development) of software applications.
- *Data consumer:* this group of users are involved in consuming data that is described using ontologies.

- *Data publishers:* this group of users are involved in publishing the structured data that is described using ontologies on the Web and in the knowledge base.

  Brief descriptions of each phase of the ODDLC are as follows:

- *Specification*: In this phase, the need or business case for a particular ontology is discussed by domain experts, users and ontology developers. The need for a particular ontology is justified by the need to perform certain business functions. Concepts like scenario analysis, use cases and scenario testing can be used to justify the requirement for an ontology with specific functions. In performing a scenario analysis, domain experts and ontology developers also identify the specifications required by the user to achieve the intended tasks.
- *Conceptualization*: In this phase, the requirements gathered in the previous phase are represented by concepts and their mutual relationships. The aim of this step is to confirm the user requirements captured by the domain experts and ontology developers using these concepts. If ontologies already exist that can be re-used, ontology developers can adapt and use them to develop the conceptualization.
- *Implementation*: In this phase, the formally agreed-upon concepts are developed by ontology developers. The desired target representation language for performing the required specifications is used by the ontology developers to progress the development of the ontology to the testing phase.
- *Integration*: This is a multi-step phase in which the developed ontology from different modules are merged together. Other steps such as testing the ontology for the suitability and desirability with respect to the application are carried out before applying it in the real world.
- *Maintenance*: In this phase, the ontology is maintained by an evolutionary process according to changing requirements or reuse.

The above-mentioned phases in the ODDLC can be classified into two broad stages of the Ontology Lifecycle namely engineering and usage as discussed in the next section.

1.5 Different Stages in the Ontology Development and Deployment Lifecycle

From a broader and wider perspective, ontologies over ODDLC steps shown in Fig. 1.8 go through two main stages in their lifecycle, namely *the engineering stage* and *the usage stage*, as shown in Fig. 1.9. The engineering stage encompasses the processes and activities involved in the construction of ontologies, while the usage stage represents the phase in which ontologies are deployed and used in the real world. The engineering stage (which is also referred to as the development stage in this book) deals with *the knowledge meta-process* [241] and focuses on knowledge identification, which includes all the relevant activities involved in the construction of ontologies such as design, implementation, evaluation and evolution (left part of

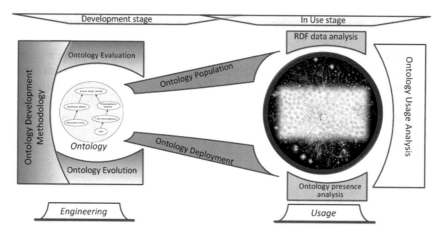

Fig. 1.9 Two main stages in ontology development and eployment lifecycle: development and in-use stages

Fig. 1.9). The usage stage (which is also referred to as the in-use stage in this book), deals with knowledge creation, which includes ontology population and the usage of the ontology (right part of the Fig. 1.9). Each stage comprises of many different steps that are explained in the following subsections.

1.5.1 Ontology Engineering

Ontological Engineering encompasses those activities that concern the ontological development process as well as the methodologies, tools and languages required for building ontologies [64]. In the literature, numerous development methodologies focusing on multiple aspects have been proposed. For example, Uschold and Kings methodology [272], METHONTOLOGY [9], and On-To-Knowledge (OTK) [249] methodologies are proposed to assist ontology developers in developing new ontologies from scratch. KACTUS [231] and the Integration-Oriented methodology [178] enable ontology engineers to reuse existing ontologies to develop new ontologies, and CO4 [92] and NeOn Methodology [221] support the collaborative and distributed construction of ontologies. A brief discussion on some of these methodologies features in the next chapter.

1.5.2 Ontology Evaluation

Since ontologies explicitly represent domains in the form of entities, properties, and relationships that exist in the real world and constitute the domain in focus, it is a practical requirement to evaluate the developed ontologies to see whether they are fit for purpose. **Ontology evaluation** is the area which focuses on measuring the quality of developed ontologies. There are several approaches to the evaluation of ontologies; for example, one might measure formal properties such as consistency and completeness,while another might look at the coverage and scope of the ontology, and yet another perspective might be to map specific upper ontologies. Functionally, ontology evaluation includes aspects of ontology validation and verification which cover structural, functional and usability issues [208].

1.5.3   Ontology Population/Ontology Deployment

Once an ontology has been developed and evaluated, it is moved to the in-use stage, as shown in Fig. 1.9 (which can also be viewed as the run-time phase) with the help of bootstrapping activities such as Ontology Population and Ontology Deployment. **Ontology Population** [4] refers to the set of activities which use automatic [112] or semi-automatic [54] techniques to populate ontologies with instance data, whereas **Ontology Deployment** refers to the set of informal techniques often used by data publishers to make use of ontologies such as Semantic Annotation [214] and Web Forms [256] to populate ontologies.

1.5.4   Ontology Evolution

Developed ontologies are meant to evolve. Changes in ontologies, as described by [204], can be triggered by three possible elements: (a) change in the domain; (b) change in conceptualization; or (c) change in formal specification (for example, change in RDF/RDFS/OWL specifications). Changes in ontologies are the focus of the **ontology evolution** research area. Ontology evolution is described as the activity of adapting the ontology to new knowledge that occurs as a result of domain changes, while preserving its consistency [290]. Ontology evolution, in general, encompasses relevant processes such as data validation, ontology changes, evolution validation and evolution management to implement the complete ontology change management process.

1.5.5   What Part of ODDLC Has Not Been Focused Upon?

There is extensive work in the literature concerning the development stage of ontologies covering ontology development, evaluation and evolution; however, less work on analysing their usage has been conducted. As mentioned earlier, analysing the

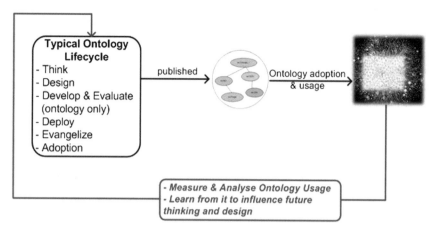

Fig. 1.10 Ontology lifecycle with a feedback loop based on ontology usage

usage of ontologies is an important step in improving the realization of ontologies and increasing their effectiveness.

Ontologies are the backbone of the Semantic Web and for them to remain useable, they need to be kept up-to-date. Ontology evolution approaches proposed in the literature have focused more on syntactical and logical aspects of ontologies to ensure their validity and consistency in their conceptual model [289]. While these aspects are important in terms of providing tools and techniques for incorporating changes (change management) in the knowledge conceptualized by the ontology, they do not provide any assistance to ontology developers and knowledge experts in obtaining feedback on how effective and beneficial (if at all) the existing implementations are. As shown in Figs. 1.9 and 1.10, such feedback is of paramount importance to the ontology lifecycle and provides pragmatic input to the different steps of engineering and usage in order to evolve a product (ontology) which is closely aligned with its users. To obtain such a feedback loop for knowledge change (ontology evolution) at the in-use stage - where instantiation by different users is experienced - a different set of techniques to evaluate and measure how the ontology is actually being used is required. This intermediary point in the ontology lifecycle, in which a set of activities is employed to analyse the ontologies while in-use, is described as **Ontology Usage** as shown in Fig. 1.9, and the analysis activity is called **Ontology Usage Analysis (OUA)**.

1.6   Why Do We Need Ontology Usage Analysis?

As discussed in Sect. 1.1, the vision of the Semantic Web is to provide a standard means for publishing data on the Web that is identifiable, accessible and under-standable by machines as well as humans. Semantic Web standards provide the foundational arrangements for the proliferation of RDF data which is semantically rich and enables data interoperability at a syntactic and semantic level. With the acceptance of Linked Data principles [140] as the best approach for publishing and interlinking structured data on the Web, we are observing a continuous growth in the publication of Semantic Web data, often dubbed Web-of-Data. Ontologies, as the formal and standard means of adding semantics to data, are also becoming widely adopted [8]. According to `PingTheSemanticWeb.com` which maintains a list of namespaces used in RDF documents, there are around 1965 namespaces (URIs) of ontologies (vocabularies) being used on the Web.[20] Another source referred to in the literature, though not considered up-to-date, is Swoogle [79] which automatically crawls the Web and has an index comprising approximately 10, 000 ontologies. As discussed in earlier sections, ontologies are being adopted in many domains, such as the Healthcare and Life Science domain, Gene Ontology [6] which is widely used to semantically describe gene-related data, Music Ontology [223] which provides a formal framework for describing music-related information on the Semantic Web, the FOAF [47] ontology which describes people, their interests and social network-ing aspects, and GoodRelations [142] which is being adopted as a vocabulary to semantically describe business entities, offers and products.

Although ontologies are being adopted in various domains, research has shown that the rate of *adoption of ontologies is not occurring at the same pace at which ontologies are being developed*. This was highlighted by [161] who conducted a study on the LOD dataset and the ontologies it contains. They state that "linked data in merely more data because of the limited use of ontologies in LOD". Other authors, too, have raised concerns about issues which impede the use of existing ontologies and the reasons for this; for example [69] suggests that the difficulty in finding a relevant ontology is the main factor hindering the adoption of ontologies. The cause of this problem can be related to the fact that application developers or practitioners do not know whether there is an existing ontology that fits their requirements, in spite of numerous ontologies having been developed. This is because there is currently no analysis that demonstrates how existing ontologies are being used.

The fact that greater advantage is being taken of domain-specific ontologies is encouraging, but for this to continue, users need to be empowered with the knowl-edge they need to facilitate the adoption and reusability of ontologies. This includes providing data publishers with information about the current ontology uptake status and the trends being observed in knowledge and data patterns. Similarly, ontology developers and domain experts need to be made aware of the variations in domain conceptualization so that ontologies can be adopted by either specializing or gener-alizing the respective concepts. However, **there is currently no formal approach in the literature to evaluate, measure, and analyse the use of ontologies on the Web** in order to provide the required visibility described above. The lack of methodical approach to performing empirical analysis on the use of ontologies will negatively

---

[20]Note: the web site PingTheSemanticWeb.com has not been live since mid-2013, but the latest statistics, captured in August 2012, are shown at https://drive.google.com/file/d/0B6jAK1TTtaSZbkp6WU5hbzNPWk0/edit?usp=sharing; retr: 10/09/2017.

impact the effective and efficient utilization of Semantic Web (RDF) data made available on the Web.

This is important considering the fact that large internet companies such as Google, Microsoft and Yahoo, after realizing the benefits of explicit semantics, have begun to support Semantic Web standards as well as Web ontologies with reasonable adoption and maturity (for example, the GoodRelations ontology [142]). As a result, billions of RDF triples published on the Web (either as part of the LOD cloud or embedded within Web documents using RDFa) and thousands of ontologies will be used to annotate the data. Having an insight into how ontologies are used will assist such endeavours.

Furthermore, ontologies are developed, published and instantiated to describe information and enable information interoperability among diverse applications. It is desirable, and also recommended by the Linked Data community, to encourage the reuse of terms defined in existing vocabularies/ontologies (where possible) to provide coherent and consistent terminological knowledge to make them more data integrated and consumer friendly. For example, a Semantic Web application can perform a simple RDF query to retrieve all the relevant data when consistent terminology is used to describe the information and map similar concepts. The reuse of terms (in RDF, this means reusing the same URIs), particularly of classes and properties, represents as the ideal situation in which highly reused concepts and properties become a de facto standard for the given domain [146]. In a given domain, once an ontology is accepted by the community, other data publishers are encouraged to reuse the adopted ontologies, which results in positive network effects. As highlighted in [141], the positive network effects that come with wider usage mean that the perceived utility of ontologies increases with the number of people who commit to ontology adoption. This signifies the importance of ontology reuse and indicates that a better understanding of **how ontologies are being used and exactly what is being used** is required to encourage the adoption of ontologies by the community.

Current information regarding the use of ontologies is limited to knowledge of which ontologies exist and how they can be accessed. The detailed ontology usage insight that can be achieved by Ontology Usage is therefore highly desirable, the potential benefit of which is described in the next section.

1.7   How will Ontology Usage Analysis help?

To describe how Ontology usage analysis will be of benefit, let us consider Bob, a user who belongs to each of the stakeholder group in ODDLC mentioned in Sect. 1.4 in three different scenarios.

1.7.1   Use Case Scenario 1: Bob as an Ontology Developer

Bob has developed an ontology, parts of which are very frequently used by consumers to describe their products. Bob is in the process of evolving his ontology, but first he wants to know which parts of his ontology need to be updated or deleted or improved. Bob wants to determine

(a) The usage of his ontology, which will give him information on how many consumers are currently using it to annotate their products.
(b) Which parts of his ontology are used the most and which are not used at all; hence, which parts need to be evolved and how. This information is necessary for Bob to ascertain whether his ontology is well regarded among consumers, and if it is, which parts are best-regarded.
(c) How the entities of a given concept are described in an ontology.

   In this scenario, Bob is generally interested in knowing how a particular ontology is being used or adopted by end users. This requires the return of a number of data sources (ontology users) which have used the ontology components (at least one term of the ontology) to describe the information published on the Web. The availability of such information reduces the likelihood of developing redundant conceptual models of a domain. It also helps to identify gaps in existing conceptual models and the need for additional coverage to make models more useful for practical implementation, which provides the key insight to ontology change or the evolutionary process.

1.7.2   Use Case Scenario 2: Bob as a Data Consumer

Bob has been approached by a business to develop an ontology that describes their product offerings, entities and descriptions. To develop an ontology according to the requirements, Bob can possibly reuse parts of numerous available ontologies, but he wants to choose the one which is most suitable and is commonly used to describe similar products. Bob therefore wants to:

(a) Obtain a list of ontologies which are currently being used by other publishers to describe similar information. This is necessary for Bob to ascertain whether a particular ontology is well-regarded among other developers in the representation of their products.
(b) Obtain information about which terms from which ontologies are the most popular and used by developers to describe similar concepts.

   To achieve this type of analysis, it is important for Bob to be aware of the knowledge patterns that are prominent and prevalent in a particular domain or application area. Knowledge patterns provide a synopsis of the occurrence frequency of terminological information in the knowledge base that describes entities or points of interest.

1.7.3   Use Case Scenario 3: Bob as a Data Publisher

Continuing the previous example, the business that has approached Bob is releasing a new product and is competing with another company for the bigger share of the market. It wants Bob to publish the data using an ontology that can be picked up by search engines when potential customers are searching for a similar product. In this scenario, Bob wants to:

(a) Obtain a list of ontologies that are currently being used to describe similar products.
(b) Obtain a list of key terms from different ontologies which are recognized by different search engines to improve the visibility of the clients product.

The ability to analyse the usage of existing ontologies would enable Bob to obtain the information he needs in each of the scenarios outlined. In connected economies and systems, we often deal with multi-disciplinary systems, and it is often necessary to support the multi-domain representation of ontologies. For example, an e-Commerce application might need to cover offering, product master data, catalogue, delivery, payment, warranty, location, legal, shipment, and order fulfilment. Therefore, it is practical and desirable to understand which ontologies maintain a cohesive grouping to provide an extended and comprehensive coverage of assimilated results from different but relevant domains. The availability of such information is essential for developing large enterprise applications based on data that is structured and semantically annotated, and can conduct knowledge extraction through reasoning.

In this book, we will propose a framework that will assist users like Bob to perform and implement such analyses to enable the efficient and effective representation and utilization of the Semantic Web.

We will investigate two problems related to the usage of ontology and believe that their resolution will, directly or indirectly, enable data interoperability and data integration on the Web.

1.8   Recapitulation

The Web is being transformed from a Web-of-Documents to a Web-of-Data. This transformation is being enabled by Semantic Web technologies which promote data interoperability through the use of ontologies. In this chapter, the role of ontologies in the realization of the Semantic Web vision was highlighted. With the continuous rise in the use of ontologies and the proliferation of Semantic Web data, the need for a solution to understand the "usage" of ontologies was highlighted and justified. In the next chapter, a survey of the literature relevant to the work of ontology usage analysis is discussed before the loop in the Ontology Development and Deployment engineering lifecycle is closed by positioning Ontology Usage Analysis as a key component.

Chapter 2
Closing the Loop: Placing Ontology Usage Analysis in the Ontology Development and Deployment Lifecycle

2.1 Introduction

The aim of this book is to present a framework to measure and analyse the *usage* of *ontologies*. **Usage** and **ontology** are the two key words that signify the focus of this work. Our understanding of these terms is as follows:

"Ontology" is an engineering artefact produced by using appropriate development methodology which comes under the definition of ontology engineering/development.

"Usage" of an ontology is an orthogonal process to ontology development and refers to situations or scenarios in which ontology is used for knowledge creation and representation.

The knowledge creation and representation process essentially relates to the instantiation of ontologies, where terminological knowledge defined by the ontology is used to semantically describe the instance data, that is encoded in the RDF data model (also known as Semantic Web data and/or Web-of-Data).

In the previous chapter, we presented the difference between the various stages of the Ontology Development and Deployment Lifecycle, from hereon referred as the Ontology Lifecycle and highlighted the importance of *Ontology Usage* in it from a practical business viewpoint. However, to support any business initiative, there needs to be sufficient backing from the theoretical perspective too in the literature. To provide a sufficiently broad background and pertinent literature synopsis from a theoretical perspective, in this chapter[1] we conduct a comprehensive survey of the literature which focuses on describing how ontology engineering (to describe

---

[1]Parts of this chapter have been republished with permission from:

1. Jamshaid Ashraf, Omar Khadeer Hussain Farookh Khadeer Hussain, A Framework for Measuring Ontology Usage on the Web, The Computer Journal, 2013, Volume 56, Issue 9, pp. 1083–1101, by permission of OUP.

2. Knowledge-Based Systems Volume 80, Jamshaid Ashraf, Elizabeth Chang, Omar Khadeer Hussain, Farookh Khadeer Hussain, Ontology usage analysis in the ontology lifecycle: A state-of-the-art review, pp. 34–47, 2017.

© Springer International Publishing AG 2018
J. Ashraf et al., *Measuring and Analysing the Use of Ontologies*, Studies in Computational Intelligence 767, https://doi.org/10.1007/978-3-319-75681-3_2

"ontology development" focused research work) and RDF data analysis (to discuss "usage" focused research work) are currently studied in the literature and applied in the real world. The chapter's discussion is divided into these two categories to delineate the existing work based on its primary focus and highlight the gap in them with respect to our proposed definition of Ontology Usage Analysis. The structure of this chapter is as follows:

- Section 2.2 presents a brief history of ontologies and its lifecycle.
- Section 2.3 presents ontology development focused work which includes:

  - a discussion of the ontology engineering discipline (Sect. 2.3.1)
  - an analysis of different ontology evaluation frameworks and the use of instance data in evaluation (Sect. 2.3.2).

- Section 2.4 presents usage focused RDF data focused work which includes:

  - a discussion of work that performs empirical analyses of RDF data on the Web (Sect. 2.4.1)
  - a discussion of analysis work that evaluates the presence and use of different ontologies and vocabularies (Sect. 2.4.2).

- Section 2.5 gives an integrated critical view of the current state of ontology and RDF usage analysis and highlights what is missing in the ontology lifecycle with respect to the Ontology Usage Analysis viewpoint discussed in Chap. 1.
- Section 2.6 positions Ontology Usage Analysis in the ontology lifecycle.

2.2 Ontology-Its History and Lifecycle

The word *Ontology* can be viewed in two different ways depending on whether one's interest is in its philosophical root or its application in computer science. In this book, we are interested in its role in the context of computer science. In the latter, ontology is written in lower case, whereas in the philosophical world, Ontology is used.

The use of ontology in computer science started around 1991 at Defense Advanced Research Projects Agency (DARPA) as part of the Knowledge Sharing Effort [199]. As is obvious from the name, the aim of this project was to find ways to develop a knowledge-based system in which knowledge is represented and reused in component form [64]. Since then, *Ontology Engineering* as a discipline has matured and provides an extensive body of knowledge to assist in the development process of ontologies. Ontologies have now become an important component of a large number of applications in many areas, including knowledge management, customer relationship management, eCommerce, biomedical, health care, data integration and eLearning, to name a few. Building an ontology requires different activities to be completed over the lifecycle. It is important to understand the ontology lifecycle from a high level perspective to group the related set of activities and generalize the lifecycle stages. Understanding the ontology lifecycle helps in identifying the

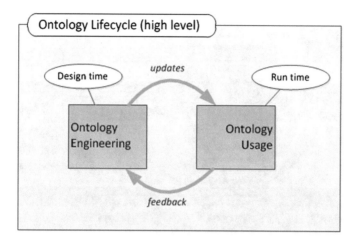

Fig. 2.1 A high level view of the ontology lifecycle

stages through which an ontology passes from its inception to its utilization, either in knowledge-driven applications or on the Web for information annotation.

According to the Oxford English Dictionary, the generic definition of lifecycle is *"a course or evolution from a beginning, through development and productivity, to decay or ending"*. In the context of ontologies, the ontology lifecycle is considered to be different from the ontology development lifecycle model. This difference has emerged from the very fact that ontology (specifically in the Semantic Web) contains the formalized representation of domain knowledge (statements expressed using OWL) but at the same time, it can contain RDF statements using the terms defined by the ontology. In other words, any document or set of statements can contain the statements describing the terminological knowledge (T-Box) and/or the statements describing instance data (A-Box). Therefore, in this book, the document which describes the conceptualized domain model is considered to be the formalized representation of the ontology. Thus, ontology lifecycle refers to the evolution through which the ontological model passes during its different stages from its development to usage. These different stages are broadly represented over the two broad high level phases of "design time" and "run-time" as shown in Fig. 2.1. "Design time" is the phase in which the ontology is being developed, and "run-time" is the phase in which the ontology is used in either a knowledge-driven application or for annotation.

Our focus in this book is on the ontology usage phase in which we want to analyze how ontologies have been received and treated after their development. There is limited focus in the literature on this area. Ontology lifecycle as a combination of the two broad phases is discussed in the literature as part of ontology development methodology, such as Cyc [88], Uschold and King [272], METHONTOLOGY [9] and On-To-Knowledge [249]. The lifecycle model discussed in these methodologies primarily relates to the models found in software engineering disciplines such as

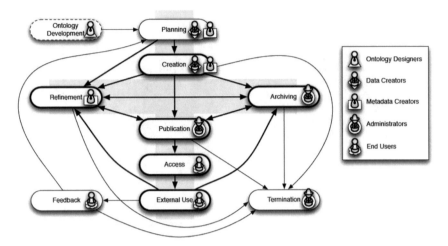

Fig. 2.2 Abstract data lifecycle model [198]

waterfall [232], spiral [34] and prototypical [5]. Another work that places emphasis on the need to analyze the usage aspect is [198]. This work surveys the lifecycle models of data and knowledge-centric systems. It first describes lifecycle models used in data-centric domains, such as digital libraries, multimedia, eLearning, knowledge and Web content management and ontology development. Based on the comparative analysis of the existing models, a meta-vocabulary of lifecycle models for data-centric systems is proposed. Using the meta-vocabulary, the Abstract Data Lifecycle Model (ADLM) is developed, along with additional actor features and generic features of data and metadata, as shown in Fig. 2.2. ADLM, being the meta-model used to describe the related but different lifecycle models, provides a comprehensive coverage of the usage aspect of ontologies (run-time stage of the ontology lifecycle; shown in Fig. 2.1). A number of the processes of the ADLM model are focused on the run-time activities through which ontologies pass while in use.

Another work [262] discusses the role of the ontology lifecycle in ontology-based information systems (OIS). In relation to the management of the ontology lifecycle, the author proposed a simplified lifecycle model, as shown in Fig. 2.3. In this lifecycle model, in contrast to those published under the label of methodologies of ontology development, equal emphasis is given to ontology usage and ontology engineering. After reviewing the different methodologies in ontology engineering activities, the author proposed that the main ontology engineering lifecycle activities were requirement analysis, development, evaluation, and maintenance. Ontology usage encompasses all the activities performed on the ontology after it has been developed and is in use i.e. in run-time state. In relation to ontology usage, all the services and processes which are involved in accessing and manipulating an ontology, such as search, retrieval and cleansing, were covered in the proposed lifecycle model. A reasoning service was also included to infer implicit knowledge and assist with expanding and

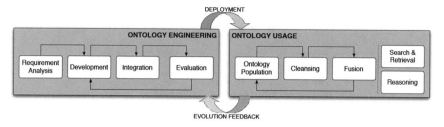

Fig. 2.3 Lifecycle model (c.f. [262])

refining the query and expanding the search results, including statements deductible through inferencing.

2.3 Ontology Development Focused Work

Immediately after the emergence of the Semantic Web [25], the significance and importance of ontologies came to the fore as an ideal approach for knowledge representation and knowledge sharing on the Web. This is the reason we see most of the work done during the early days of the Semantic Web (from 1999–2006) was centred around ontologies and different methodologies to develop and support them. This work included:

- methodologies and frameworks for developing ontologies under the name of *Ontology Engineering* [162, 197, 251])
- formal languages to represent ontologies under the name of *Ontology Languages* [156, 192]
- methodologies to evaluate ontologies under the name of *Ontology Evaluation* [40, 45, 104, 257]
- methodologies to evolve ontologies under the name of *Ontology Evolution* [204, 280]
- formal logic for reasoning with ontologies under the name of *RDFS/OWL Reasoning* [73, 194, 236].

Each of the above-mentioned areas is the subject of individual research effort; however, the research community groups them under the rubric of Ontology Engineering. Ontology engineering generally covers two sets of activities; (a) ontology development, methodologies and processes; and (b) tools and languages for supporting and automating ontological development as shown in Fig. 2.4. These are discussed further in the next sub-sections.

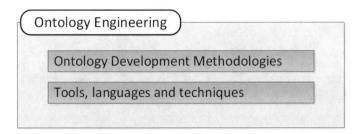

Fig. 2.4 Ontology engineering components

2.3.1 Ontology Development Methodologies and Processes

Similar to software engineering and software development lifecycle models [35], ontologies are developed and maintained using ontology development methodologies which are important components of Ontology Engineering [115], as mentioned in Sect. 1.4. Most of the present methodologies, such as On-To-Knowledge Management [253] and METHONTOLOGY [98], tend to cover the engineering aspects of the lifecycle which include *requirement analysis*, *ontology development*, *evaluation* and *maintenance* [262].

Before the different methodologies, and the methods and processes proposed, developed and deployed for the development of ontologies are discussed, the definition of methodologies and methods standardised by the IEEE Standard Glossary of Software Engineering Terminology [222] are reviewed next.

Methodology: Methodology is a comprehensive, integrated series of techniques or methods creating a general systems theory of how a class of thought-intensive work ought to be performed.

Method: A method is an orderly process or procedure used in the engineering of a product or in performing a service.

Technique: A technique is a technical and managerial procedure used to achieve a given objective.

Process: Process is a sequence of interdependent and linked procedures which, at every stage, consumes one or more resources to convert inputs into outputs.

A growing number of methodologies are proposed for ontology development. Few methodologies attempt to cover the whole lifecycle of ontology engineering, ranging from requirement elicitation and development to evaluation and maintenance, while others focus on a specific stage or process of ontology engineering. Over two dozen methodologies and methods supporting ontology development from the literature are presented in chronological order in the next sub-section. Among them, few which are relevant to the scope of this book will be discussed in some detail. The methodologies and methods have been categorized according to the following classifications:

- methodologies which develop ontologies from scratch (discussed in Sect. 2.3.1.1)
- methodologies which support the cooperative and distributed construction of ontologies (discussed in Sect. 2.3.1.2)
- methodologies which use Web 2.0 features to provide social networking aspects in ontology development (discussed in Sect. 2.3.1.3)

Each of these groups is discussed in subsequent subsections.

2.3.1.1   Methodologies and Methods for Building Ontologies from Scratch

Methodologies which support the creation of ontologies from scratch are briefly described as follows:

- Cyc [88, 177]: Cyc methodology was the result of experience gained through the development of the Cyc knowledge base comprised of common sense knowledge. A detailed description of Cyc methodology and knowledge bases is available at www.cyc.com (retr.; 11/11/2017).
- Uschold and King's methodology [268, 270, 272]: This methodology is the result of research conducted on the development of Enterprise Ontology to model enterprise processes. This ontology represents the terms and definitions relevant to business enterprises. The detail on Enterprise ontology and the methodology can be accessed from http://www.aiai.ed.ac.uk/project/enterprise/ (retr.; 11/11/2017).
- TOrinto Virtual Enterprise Methodology (TOVE) [123, 124, 270]: In the literature, this is also known as Gruninger and Fox's methodology. As a part of the TOVE project, the methodology comprises several steps: (a) motivation scenarios; (b) informal competency questions; (c) first-order logic-based terminology; (d) formalization of competency questions; and (e) definitions of semantics and constraints. One of the significant elements of this work is that it is considered to be the first reported use of competency questions in defining the scope of ontology. Further detail on TOVE can be found at http://www.eil.utoronto.ca/enterprise-modelling/entmethod/index.html (retr.; 13/11/2017).
- KACTUS [230, 231, 287]: This is a European ESPRIT-III project aimed at evaluating the feasibility of knowledge reuse in complex technical systems and the role of ontologies in giving explicit structure to the knowledge. Using the methodology, the authors developed three ontologies and applications: fault diagnosis in electrical networks, scheduling service resumption after a fault appears, and control of electrical networks.
- METHONTOLOGY [9, 98, 180, 275]: This is one of the most famous ontology development methodologies. It defines a comprehensive set of activities needed for the development and maintenance of ontologies. The authors also describe the lifecycle of an ontology, from requirement gathering to the evolution of the ontology. The stages through which the ontology passes are: *specification, conceptualization, formalization, integration*, and *implementation*. In addition to these core development centric activities, a number of umbrella activities such as *evaluation* and *documents* are used which run through the lifecycle stages. One of

the significant achievements of METHONTOLOGY is that it has been selected for consideration for the development of ontologies by the Foundation for Intelligent Physical Agents (FIPA),[2] which promotes communication across agent-based applications. Further detail on METHONTOLOGY can be found at: http://www.oeg-upm.net/ (retr.; 12/11/2017).

- SENSUS [165, 166, 254, 273]: This is one of the early approaches to knowledge sharing using ontologies. This approach is based on the assumption that if two knowledge bases are using the same base ontology, then knowledge can be easily shared between them since they share a common structure. The SENSUS methodology comprises the following steps: (a) a list of terms are identified as seed terms that are particular to the domain; (b) seed terms are linked with the SENSUS ontology (the SENSUS method makes use of the SENSUS ontology which has more than 70,000 concepts organized in hierarchy according to their abstraction level [97]); (c) all the concepts in the path from the seed terms to the root of SENSUS are then included; (d) relevant terms which are missing are added manually; and (e) lastly, for those nodes with a high betweenness value, the entire sub-tree under this node is added. Using this approach, an ontology for military air campaign planning was built which describes basic elements such as the air campaign plan, scenarios, commanders, and participants [273]. Further detail on methodologies and ontologies can be found at http://www.isi.edu/natural-language/projects/ONTOLOGIES.html (retr.; 17/11/2017).

- On-To-Knowledge (OTK) [249–251, 253]: Part of the EU IST-1999–10132 project, the On-To-Knowledge (OTK) methodology was developed for the introduction and maintenance of knowledge-based applications in enterprises focused on knowledge processes and knowledge meta processes, based on ontologies. This methodology comprises the following stages: (a) *kick-off*: requirements are identified, competency questions are identified and the final draft of the ontology is developed either from scratch or reusing existing ontologies, (b) *refinement*: the ontology is refined to meet the application requirements; (c) *evaluation*: the ontology is evaluated using competency questions to measure its usefulness; and (d) *ontology maintenance* the ontology is updated to reflect changes. This project was later joined by the Ontotext company[3] in 2001 to develop ontology middleware and a reasoning module based on the work that went into the On-To-Methodology. Ontology middleware developed through this collaboration [49] provided the administrative layer on top of On-To-Knowledge to make this research work more integratable with real-world applications. On-To-Knowledge methodology details are available at http://www.ontotext.com/ and http://www.aifb.kit.edu/web/On-To-Knowledge/en (retr.; 26/10/2017).

- DOLCE [60, 207, 246]: DOLCE stands for a Descriptive Ontology for Linguistic and Cognitive Engineering. The main idea behind this project was to develop first-order logic-based ontologies for inclusion in the WonderWeb foundation Ontologies Library [154]. DOLCE, an upper level ontology, was the first mod-

---

[2]http://www.fipa.org; retr.; 12/11/2017.

[3]http://www.ontotext.com/ retr.; 05/11/2017.

ule of this library to be built by introducing the concepts informally along with the basic categories, functions and relations initially. Later, detailed axiomatization was added to impose the constraints on the model and clarify the assumptions through the illustration of formal consequences [189]. As part of this project, the KAON open-source ontology management infrastructure [278] was also developed to provide tools to manage ontologies. It includes a comprehensive tool suite allowing easy ontology creation and management, as well as the construction of ontology-based applications [154]. Further detail on the WonderWeb project, and its deliverables can be found at http://www.istc.cnr.it/project/wonderweb-ontology-infrastructure-semantic-web (retr.; 13/11/2017) and for DOLCE visit http://www.loa.istc.cnr.it/old/DOLCE.html (retr.; 21/11/2017) respectively.

- KBSI IDEF5 [18, 121]: IDEF5, which stands for Integrated Definition for the Ontology Description Capture Method, is an ontology engineering approach to the development, modification and maintenance of domain ontologies. The IDEF5 method is part of the IDEF family of modeling languages in the field of ontology engineering. This method considers ontology development as open-ended work which cannot be effectively adopted using a "cookbook" approach; therefore, the authors published a general procedure with a set of guidelines comprising the following activities:

  - *organizing and scoping*: the purpose, viewpoint and context for the ontology development project is identified and roles are assigned to team members.
  - *data collection*: the raw data required for the development is gathered using typical knowledge acquisition techniques, such as protocol analysis and expert interviews.
  - *initial ontology development*: the data obtained from the previous activity is used to build a prototypical ontology which contains proto-concepts (concepts, relations and properties).
  - *ontology refinement and validation*: the proto-concepts are iteratively refined and tested. This is essentially a deductive validation procedure to refine and validate the ontology to complete the development process.

  According to the IDEF5 methodology, the initial ontology is defined with a schematic language which is a set of graphical notations used to express the most common form of ontological information. Further details on the IDEF5 methodology and the IDEF5 schematic language can be found at http://www.idef.com/IDEF5.htm (retr.; 19/11/2017).

  In addition to the above-mentioned methodologies, many approaches have been proposed to address a specific aspect of ontology development. Brief details of these methods are provided below.

- MENELAS [193, 296, 297]: MENELAS is based on four principles: similarity, specificity, opposition and unique semantic axis, which helps in the development of taxonomic knowledge in ontologies. Based on these four principles, the MENELAS ontology was designed as part of a natural language understanding system. MENELAS was then used to develop an access system for medical

records using natural language. Further details on MENELAS can be found at
http://estime.spim.jussieu.fr/Menelas/ (retr.; 20/11/2017).

- ONION [106, 108, 242]: The ONIONS (ONtological Integration Of Naive
  Sources) project was initiated in 1990 to address the problem of conceptual heterogeneity, particularly in the medical domain. The object of the project was to develop
  a large-scale ontology library for medical terminology. The terminological knowledge in this approach is acquired by conceptual analysis and ontology integration.
  The ONIONS methodology exploits a set of formalisms, a set of computational
  tools that implement and support the use of the formalisms, and a set of generic
  ontologies taken from the literature in either formal or informal status and translated or adapted to the formalism proposed by [107]. For more detail visit http://
  ksi.cpsc.ucalgary.ca/KAW/KAW96/steve/introduction.html (retr.; 29/10/2017).
- COIN [182, 183]: The COntext INterchange (COIN) strategy, developed at MIT's
  Sloan School of Management, is an approach for solving the problem of interoperability of semantically heterogeneous data sources through context media-
  tion. It provides the notations and syntax to represent an ontology. This approach
  attempts to resolve semantic conflicts between heterogenous systems by defining
  the context axioms corresponding to the systems involved in the interaction. It also
  provides formal characterization and reasoning underlying the context interchange
  strategy [113].

In the next subsection, brief overview of ontology development methodologies
that support cooperative and distributed approaches is presented.

2.3.1.2   Cooperative and Distributed Approaches for Building Ontologies

An ontology is a shared and common understanding of a domain which is built
by establishing an agreement among domain experts on the conceptual model of
the domain. Since ontologies have to be available on the Web and the end users of
the ontologies may be from different locations, it is important to have methodologies which support the distributed development of an ontology in order to arrive at
consensus on the ontological model, as shown in Fig. 2.5.

Selective methodologies which support the distributed and collaborative ontology
development process are briefly described below.

- CO4 [91–93]: Collaborative Ontology (CO4) is one of the earliest works started
  at INRIA[4] toward developing ontologies cooperatively. It enables different par-
  ties to discuss, share and establish agreement on the domain model to represent
  consensual knowledge in the knowledge base. Consensus on the content of the
  knowledge base is achieved by a protocol which integrates knowledge, based on
  that consensus. The knowledge base architecture is shown in Fig. 2.6.

---

[4]http://www.inria.fr/; retr. 23/10/2017.

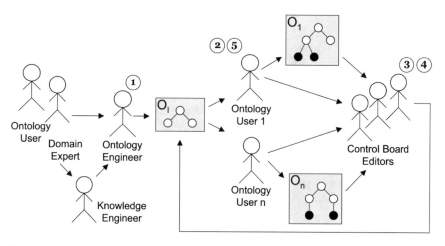

Fig. 2.5 Roles and functions in distributed ontology engineering (cf. [282])

- NeOn Methodology [32, 221, 248]: NeOn which takes its name form Network Ontologies is the latest methodology to support the collaborative aspects of ontology development, reuse and evolution in distributed environments. It is considered to be a flexible, scenario-based methodology for building ontology networks which provides variety of pathways for ontology development [248]. The key components of the NeOn methodology are: (a) a set of nine commonly occurring scenarios for building ontologies, such as when to re-engineer the available ontology, and align, modularize and localize this with ontology design patterns; (b) the NeOn glossary of processes and activities; and (c) methodology guidelines for each process with (i) a filling card, (ii) a workflow, and (iii) examples. The nine possible scenarios and the expected output and existing knowledge resources to be reused are shown in Fig. 2.7. Further details on the NeOn methodology are available at: http://mayor2.dia.fi.upm.es/oeg-upm/index.php/en/methodologies/59-neon-methodology/ (retr.; 20/10/2017).
- HCOME [172–174]: The HCOME methodology is a human-centred approach which integrates argumentation and ontology engineering in a distributed setting, where ontologies are considered living artefacts, ontology development is considered a dynamic process and special focus is placed on ontology evolution throughout the ontology lifecycle. In HCOME, expert users formalize their own ontology first before sharing with others (through the Shared Space) to evolve the conceptual model. After deliberation by the domain-specific community, agreement is achieved before moving the ontology into the Agreed Space.
- MeltingPoint [109]: The Melting Point (MP) methodology provides a collaborative ontology development environment in decentralized settings. MP methodologies are the result of experience obtained by the authors through their work in the biology domain. This methodology reuses some components of several ontologies and analyses their reusability in the MP methodology.

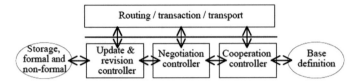

Fig. 2.6 The software architecture. Each box represents a software module, each circled unit is a data/knowledge repository and each arrow represents the call of a program functionality [92]

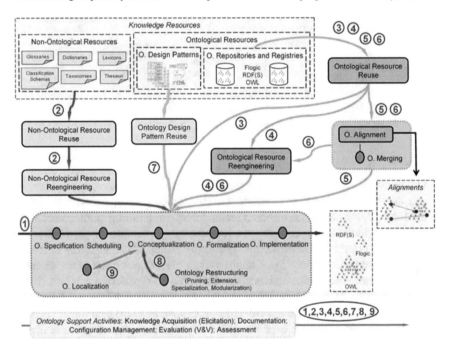

Fig. 2.7 Scenarios for building ontology networks [116]

In the next sub-section, the third category of methodologies in which Web 2.0 approaches to provide social networking aspects in ontology development are briefly described.

2.3.1.3 Ontology Development Approaches Using Web 2.0 Social Networking Features

This section describes few of the methodologies which support collaboration-based ontology development in decentralized settings. The emergence of implicit semantics based on social interaction on the Web (i.e. social web sites) has motivated ontology researchers to use Web 2.0 technologies in developing ontologies. Several techniques, such as social tagging systems (STS) [144] which allow users to freely associate terms with resources, are being used to allow users to provide implicitly conceptual structures and semantics. Such conceptual structures are known as folksonomies and are increasingly being used for information retrieval, discovery and clustering on the Web. Some of the methodologies which have considered social interaction and Web 2.0 approaches in the ontology engineering process are:

- FolksOntology [274]: A comprehensive approach for driving ontologies from folk-sonomies by integrating multiple techniques and resources is proposed. Statistical analysis techniques are used to measure folksonomics usage, structure, implicit social networks and compare them with knowledge resources such as Wikipedia, WordNet and online dictionaries. Following data analysis, ontology mapping and matching techniques are used to create correspondence between terms to develop consensus over ontology elements.
- Ontology Maturing [42]: This approach considers ontology engineering more as a collaborative informal learning process and less as a specialized knowledge engi-neering approach. Thus, in this methodology users engage in ontology engineering in their everyday work processes by integrating tagging and folksonomies with formal ontologies. This makes ontology development a learning process which is continuous and evolutionary. The development process is structured into four phases known as the knowledge maturing process: (a) emergence of ideas; (b) consolidation in communities; (c) formalization; and (d) axiomatization.

2.3.2 Ontology Evaluation Frameworks

Developed ontologies are evaluated to measure their quality and fitness based on the requirements specification. It is very important to evaluate the ontology in the development phase to ensure that when it is used, it is *fit* to serve the purpose. Several frameworks for ontology evaluation are available; however, before proceeding with discussion on different ontology evaluation approaches, we will discuss how ontology evaluation relates to the ontology lifecycle, as shown in Fig. 2.8.

Ontology Evaluation (OE), often described as a sub-area of ontology engineer-ing, covers research pertaining to the measurement of the quality, usefulness and fitness of the developed ontology, *with or without considering the instance data*. Since ontologies are an important component of the Semantic Web, ontologies can have diverse usage scenarios not known to the ontology developer. Therefore, it is important to also evaluate how a particular ontology is being received and used in the real world. There are two stages in the ontology lifecycle at which the ontology needs to be evaluated to provide a comprehensive feedback loop to all ontology stake-holders as shown in Fig. 2.8. First, *during and after ontology development* (known as Ontology Evaluation) and second, *while the ontology is in use* (known as Ontol-ogy Usage Analysis). Most of the literature covers the first evaluation point where the ontology is assessed prior to its actual utilization (Ontology Evaluation stages).

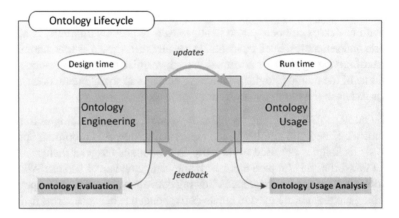

Fig. 2.8 Ontology lifecycle model with ontology evaluation

However, the difference between evaluating an ontology in Ontology Evaluation and usage analysis is explained further in Sect. 2.6.1.

Since the initial awareness of the value of ontologies, ontology evaluation frameworks have been proposed as part of ontology engineering to assess ontologies from different dimensions. One classification of ontology evaluation approaches is in [40] in which ontologies are categorized based on their objectives, as follows:

- Golden standard: where an ontology is used to evaluate other ontologies
- Application-based: using the ontology in an application and evaluating the results
- Data driven: analysing the content of an ontology to measure its domain coverage
- Assessment by humans: conducted by humans to assess whether the ontology meets the required criteria.

Other research has studied the structural aspects of ontologies to understand the relationship between ontology usefulness (fit for purpose) and its topological properties. For example, in [258], the authors presented a framework and a set of measurements to evaluate the richness, connectivity, fullness and cohesiveness of a given ontology. They proposed metrics and measures in their study to determine whether an ontology is domain-specific or generic. In [128], the authors proposed the OntoClean methodology in their approach to **evaluate** and **validate** the ontology's taxonomical relationship by employing formal notions from philosophy such as essence, identity and unity. Four meta-properties (rigidity, identity, unity and dependence) and operators $(+, -, \sim)$ to symbolically specify the characteristics of ontology components, such as classes and relationships, were used to validate the assumption which influenced the ontological model. In other research, authors [1] proposed a method for the evaluation and ranking of ontologies based on four ranking measures: class match measure (evaluates the coverage of an ontology for a given search term); density measure (measures the density based on its neighbourhood i.e. sub-classes, super-classes, relationships and siblings); semantic similarity measure (how closely the concepts of interest are laid out in the ontology); and betweenness measure (how far a concept is from the root concept of its hierarchy). Evaluation approaches [128], which either relied largely on the structural aspect of the concepts or the integrated evaluation of ontologies approach [1], provide insight on how knowledge is distributed and the existence of different thematic hierarchies in a given ontology. Recent work in this area was reported by Vrandecic in his dissertation [280] which answers the research question, *How should the quality of an ontology on the Web be assessed?* The author proposes concrete measures relating to quality, computational efficiency, accuracy, usefulness and adaptability of the ontology which he defines as a formal artefact comprising classes, properties and instance data. He argues that the right approach for ontology evaluation is to find the methods and metrics which tell us whether the ontology is flawed and if it is, in what ways, instead of merely evaluating an ontology on its positive features.

Another different but related ontology evaluation approach is **ontology evolution** which implements change management in ontology engineering. Stojanovic [243] presented the theory and practice of ontology evolution and it has been addressed by several researchers by defining change operations and change representations for ontology languages. Authors in [243] presented evolution strategies to handle inconsistencies for evolving ontologies in a centralized setting and for the handling of ontology changes in a distributed setting. However, change operations have been proposed for specific ontology languages, such as OKBC [57], OWL [219] and the KAON ontology language [39]. Based on work by [243], Haase [132] extended the work for OWL-DL ontologies which focused on consistencies while investigating the ontology evolution. The authors also developed a tool called evOWLution which implements their approach.

So, as can be seen from the discussion till now, different authors have proposed different methods to focus on specific aspects of ontology during evaluation. While some authors [70, 100, 128] have focused on functional completeness, generality, efficiency, perspicuity, precision granularity, and minimality of ontologies, others, such as [158, 292], have considered the **structural aspects** of an ontology to measure topological characteristics to infer their effectiveness and usefulness. The significant amount of literature that focuses on evaluating ontologies before they have been used highlights the absence of work that evaluates ontologies while they are being used. The assessment of ontologies in use could provide pragmatic and empirical assessment on how vocabularies are being used and adopted, which in turn could help to improve the ontological model and effective knowledge utilization.

In the following section, work in which ontologies and their usage is analysed to measure the usage, data patterns and knowledge patterns in the RDF dataset is presented.

2.4 Semantic Web (RDF) Data Focused

As mentioned in Sect. 1.6, there has been tremendous growth in the publication of structured data on the Web since the introduction of Linked Data principles [140] by Sir Tim Berners-Lee in 2006. The growth in the RDF is credited to the simplicity of the four linked data principles (mentioned in Sect. 1.2) which have provided an easy-to-follow approach for publishing Semantic Web data on the Web. Another effort which has significantly contributed to the proliferation of RDF data on the Web is the research community effort dubbed the Linked Open Data (LOD) project [169]) which has contributed billions of schema-level and instance level triples on the Web, covering myriad application areas. It is well known that Semantic Web data is loosely comprised of two types of statements: one in which terminological knowledge is described, and another in which instance data is defined. Though both types of statements are encoded and stored in the same documents, their classification helps in conducting more focused analysis depending on the objectives. In order to provide sufficient coverage on the work in which RDF data is analysed to find the data and knowledge patterns, and share best practices, the relevant work in this section has been categorized on the basis of its focus. In Sect. 2.4.1, we present the work in which *empirical analysis is performed on RDF data*, and in Sect. 2.4.2, the work that *evaluates the presence and use of different ontologies and vocabularies* is described.

2.4.1   Empirical Analysis of RDF Data on the Web

In this section, the work in which RDF data is analysed to understand the data and knowledge patterns available is described. Several research efforts have made use of the Linked Open Data (LOD) cloud datasets to perform empirical analysis, which is published as part of the ISWC (International Semantic Web Conference)'s Semantic Web Challenge.[5]

The simplicity of Linked Data principles [30], introduced by Tim Berners-Lee in 2006, and the consequent monumental success of the Linked Open Data project [140] transformed the Web into a structured data space. This new data space, comprising self-describable data based on a standard model (RDF), provided a test bed for researchers[6] to unleash and exploit the potential of Semantic Data on the Web. Researchers have analysed the Web of Data to understand the nebulous nature of the **quality of data**. One of the early attempts to analyse the quality of RDF data was made by Hogan et al. [148], who reported the common errors made by early RDF data publishers. While highlighting the shortcomings, issues and findings, the authors provided guidelines for both data publishers and consumers to assist in generating and consuming high quality semantic data. An analysis experiment was carried out on a dataset comprising data crawled from 150,000 URIs. While the prime focus was to measure noise and inconsistency in the dataset, the authors classified the errors into four categories: (a) *accessibility*; (b) *syntactical errors*; (c) *reasoning*; and (d)

---

[5]http://challenge.semanticweb.org/; retr.;14/12/2017.

[6]For example, Linked Open Data Around-The-Clock (LATC) is a European-funded project, to *"create an in-depth* **test-bed** *for data intensive applications by publishing datasets produced by the European Commission, the European Parliament, and other European institutions as Linked Data on the Web"* as one of their objectives (http://aksw.org/Projects/LATC.html; retr.; 19/12/2017).

non-authoritative contributions. One of the significant findings was that 14.3% of the properties (URIs used as a predicate in triple) and 8.1% of the classes were used in ways for which their declaration and description is not available. In addition to this, they also reported the presence of certain vocabulary terms that were used in a manner counter to their original semantics and purpose, a practice known as ontology hijacking. Their work reports on the quality of the RDF data, which is very subjective, therefore it cannot be generalized since each application has its own data requirements and specific modelling choices. Hence, it is believed that while analysing "data", one also needs to look at the traces on "knowledge" on the dataset to allow the maximum utilization of analysis findings. In another work [12], the authors have tried to **identify the shortcomings** of the LOD Cloud and suggest ways in which it could be improved and used for practical purposes. Specifically, they highlighted: (a) the need to improve the performance of RDF data management tools to provide efficient data processing; (b) the need to improve the state of "interlinking" among diverse datasets to provide a typed linked dataspace; (c) the need to improve the algorithms and tools to enhance the quality of the linked data; and (d) the need to provide an adaptive user interaction experience to support linked data management services.

In 2008, Hausenblas et al. [138] attempted to empirically **gauge the size of the Semantic Web** when there was a surge in RDF data. This was also the first work in which authors tried to study schema level data and instance level data to understand their hidden patterns. Instance level data was further classified into single-point-of-access and distributed datasets. In their analysis, the authors report on the number of triples available, the frequency of the subject, object and predicates and the level of external linkage (external linkage refers to triples in which the subject and object refer to the resource hosted in different domain names). They found that FOAF data is well connected internally and sparsely with external resources. Although they did not mention the effective size of the Semantic Web, they provided an estimate on how well the Semantic Web is linked and what types of dataset are available.

In another study, Mika et al. [196] identified the **semantic gap** which is essentially the divide between the supply of the data on the Semantic Web and the demand of a typical web user. They provided a generic method for extracting the attributes that Web users are searching for regarding particular classes of entities. Using this, they contrasted class definitions found in Semantic Web vocabularies with the attributes of objects in which users are interested. The study was conducted on data comprising different data formats, such as eRDF, RDFa data and certain popular microformats. Although they argued that RDFa is becoming more popular compared with other formats, RDFa was 0.6%, much less popular than the other formats in their dataset, particularly `hcard`, which was the most popular during 2008 and 2009. Their work found that Semantic Web technologies could play an important role in web searches if web sites published structured information to target a particular category of queries.

Aside from looking at the **data quality issues** in RDF data, Semantic Web **effective size** and **interlinking** between decentralized datasets, identifying the **semantic gap** between the available data and users' expected Web search results, several researchers also looked at the generic **characteristics** of Semantic Web data, described as follows.

Semantic Web data includes a wide range of topics such as quantifying RDF data patterns [148], analysing the distribution of schema level details in RDF documents [138], and the statistical properties of the LOD cloud [31]. The early work on **characterizing Semantic Web data** on the Web was reported in [78]. In this work, the authors estimated the number of RDF documents available on the Web, based on the search engine result pages (SERP) returned by Google. The estimated number of RDF documents found at that time (i.e. 2006) was in the range of 10^7–10^9. The authors also provided an in-depth analysis, conducted on 300M triples (mostly consisting of FOAF with some RSS1.0 documents), on the landscape of RDF web data, including the number of files, provenance in terms of website, use of RDFS primitives, and use of class and properties. They found that 2.2% of classes and properties had no definition and that 0.08% of terms had both class and property meta-usage. Other work based on a similar analysis approach but from a Linked Data perspective was reported by Hausenblas et al. in [138]. The motivation of their study was based on the argument that understanding the size of the current Semantic Web is critical to the development of scalable Semantic Web applications. Empirical analysis comprising syntactical and semantic aspects was conducted on an LOD dataset which was viewed as an interlinked (single-point-of-access) dataset and distributed (FOAF-o-sphere) datasets. For the interlinked datasets, they reported on the number of triples available and automatic interlinking, which yields a high number of semantic links, but they are of shallow quality. As such, no quantitative measure is reported by the authors to size the Semantic Web data, but this helps to illustrate the importance of creating semantic links among distributed datasets.

While, the current approaches in this category analyse RDF data from different dimensions, such as **quality**, **data patterns**, **structural properties**, **interlinking** and general **characteristics** of datasets, it is equally important to understand how vocabularies and ontologies are being used on the Web. An overview of the literature in which schema level data, which include ontologies and vocabularies, are analysed and assessed is described in the next sub-section.

2.4.2 *Empirical Analysis of Ontologies and Vocabularies in RDF Data*

In this sub-section, the reported work on the use of ontologies, including both **W3C vocabularies** (i.e. RDF, RDFS, OWL) and **domain ontologies** on the Web, is discussed.

A large amount of work has been reported on evaluating the usage of W3C-based standard vocabularies. The authors in [66] surveyed 1300 OWL ontologies and RDF schemas and reported some of the trends observed during the investigation. They observed that most of the ontologies from the OWL family are OWL DL and OWL Lite ontologies. Also, Cheng et al. [58] conducted a study on roughly

3000 vocabularies, comprising 396,023 classes and 59,868 properties in total. In addition to vocabulary documents, the authors also considered 15 million instance documents to investigate the relatedness between ontologies. They reported that 72% of vocabularies contain no more than 24 terms and also investigated the relatedness indicators between vocabularies, the textual content of vocabularies, and the explicit linking between vocabularies.

Reference [66] also reported on the characterization of the Semantic Web, based on the WATSON repository which represents a snapshot of the online semantic documents available during 2006. One of their findings is similar to the finding we reported in [8], that a large number of small and lightweight ontologies are used in the same space as large scale heavyweight ontologies. They highlighted the need for an effort to improve the quality and usefulness of existing ontologies and the need to develop ontologies for diverse domains.

Also reported in the earlier sub-section, Ding et al. [81] collected 1.5 million RDF/XML documents from the Web and reported on the use of different namespaces and the concepts and properties defined by these namespaces. Their particular emphasis was on the documents in which information (data) is semantically described with FOAF and DC vocabularies. They found that the majority of the RDF data is published by a small number of social network sites such as `livejournal.com`, `academy.com` and `deadjournal.com`. Apart from reporting that a large amount of RDF data is, in fact, published by only a small number of data publishers, they also analysed the network properties of the FOAF network such as connected components and the distribution of nodes. They detected various forms of Zipf distribution, such as the number of `foaf:Person` described in each document and the number of aliases found by using `foaf:mbox_sha1sum` predicate.

Ding et al. [78] presented another analysis conducted on 300 million RDF triples, on the use of FOAF and DC vocabularies on the Web. They described the number of global metrics, properties and usage patterns for the study and also observed the presence of the power law distribution of different metrics. They reported that most of the classes (>97%) are not instantiated on the Web (that is, not used to define instances) and likewise, more than 70% of the properties are also not used to describe resources. One of the significant contributions of their work was a discussion on whether or not the traditional monolithic ontologies are the best solution for the Web, or whether the research community should proceed with lightweight vocabularies and encourage maximal reusability of existing vocabularies.

Another important focus in analysing the schema level data on the Web is to look into the use of **standard W3C meta-vocabularies** such as RDF, RDF Schema and OWL/OWL2. Several researchers have investigated the use of the `owl:sameAs` predicate which allows two resources to refer to the same things on the Web (meaning that two co-referent resources talk about the same real-world object).

In another work, Ding et al. [80] presented work in which they explored the presence of `owl:sameAs` to combine and retrieve additional information during crawling. They reported on the quality issues observed, such as the casual use of `owl:sameAs` without giving due recognition to the fact that the symmetric semantic of `owl:sameAs` can create considerable Web discrimination.

Reference [150] looked at **co-referencing issues**, keeping in mind the OWL features that allow inferences, including inverse functional property and functional property. They explored the use of `owl:sameAs` in the same dataset by computing inference closure and found that URIs with at least one alias had an average of 2.65 aliases due to the incorrect use of `owl:sameAs` linkage. They also reported that 57% of alias groups contained URIs from multiple domains. They looked at the implicit `owl:sameAs` relations which were produced through inference over inverse-functional properties, finding that the majority of additional aliases had blank-nodes coming from the same domain. Overall, the finding was that the quality of Linked Data is high if under taken carefully; otherwise it can be a burden on the applications which consume this type of data.

In a recent work, Cheng et al. [58] performed an empirical study on a dataset collected from 261 pay-level-domains comprising 2,996 vocabularies. These vocabularies contain 396,023 classes and 59,868 properties. The authors took 15 million instance documents to investigate the relatedness between vocabularies, measured with respect to how terms are defined, the textual content of vocabularies, and co-occurrence in instance documents. They also looked at the relationship between relatedness and popularity and its usage for Falcon's ontology search recommendation service. The significant finding of their work is that several related vocabularies are not interlinked, and those which are interlinked are often co-used in the same instance document.

In the study by [8], the authors analysed the usage and adoption of the GoodRelations ontology in the eCommerce domain. To base the findings on empirical grounds, a purpose-built dataset was used containing RDF (most represented using RDFa) of the data from 105 different pay-level domains. The co-usability factor of the domain ontology with other ontologies was analysed to observe how different vocabularies are being co-used to semantically describe the entities pertinent to the eCommerce domain. Using real use cases, the use of different object properties and attributes of pivotal concepts (`gr:Offering`, `gr:BusinessEntity` and `gr:ProductOrService`) were analysed to understand the data and knowledge patterns available on the Web in eCommerce domains. One of the findings of this work is that a small proportion of the ontology is hugely used by a large number of data sources. This supports the previous findings and recommendations [78, 148] that Web ontologies, in order to be successful on the Web, should be small rather than monolithic.

2.5 What is Missing with Respect to Ontology Usage Analysis?

In this section, a critical evaluation of the existing approaches in the literature is presented and the main issues that need to be addressed for measuring ontology usage are identified. As can be seen from the discussion in previous sections, there are several approaches in the literature proposed by different researchers in which *ontologies are developed* and the *RDF data is analysed* from different perspectives. Relating to the development of ontologies, approaches have been proposed to *develop* and *evaluate ontologies*, measure their quality and assess their compliance with the requirements. Also, approaches have been proposed to *evolve ontologies* to implement change management, to ensure that ontologies remain useful and adapt to the new requirements.

However, none of the approaches reviewed conducted an analysis of ontologies based on their usage in a real world setting; this therefore brings the *using* element of ontologies to the fore. Clearly, in evaluating ontologies and analysing Semantic Web data when an ontology has been developed, there is a need to bring the "usage" aspect of ontologies and RDF data into the equation to better understand issues of adoptability and uptake in the actual instantiation. In the literature, RDF data has been analysed to assess the quality and understand the use of W3C-based vocabulary constructs in the published instance data. The approaches analysed the instance data but not from the perspective of measuring the use of different domain ontologies. Therefore, in addition to analysing the RDF data from a quality perspective, it should also be analysed from the ontology usage perspective to measure the semantic level information in the instance data. Having such information will provide pragmatic insights about the state of ontology usage and its level of adoption, and will assist better with the development of evolution strategies. In summary, the shortcomings identified in the existing literature related to ontology usage analysis are:

- Most of the ontology lifecycle models are centred on the construction and evaluation of ontologies. Here, the emphasis is on developing approaches for ontology development which closely match their anticipated "usage" and which, once developed, will be evaluated according to this factor. However, no emphasis is placed on evaluating the "usage" of the developed ontologies in real world settings from the viewpoint of their instantiation.
- Most of the ontology evaluation approaches only consider the formalized conceptual model to evaluate ontologies. Since ontology evaluation, in most methodologies, is considered part of the development phase in order to measure the effectiveness of a developed ontology, "usage" is not measured due to its lack of implementation in real world applications. Therefore, the concept of "usage" often refers to the evaluation of the use of different constructs to describe the concepts and other components of the ontologies, and not their "usage" in annotating information.
- Most of the RDF analysis work focuses on analysing the quality aspect of published RDF triples. Here, the "usage" concept is used to analyse the different W3C-based vocabularies and their compliance with the linked data principles, but it does not analyse the "use" of domain ontologies in semantically describing the information.
- Most of the work in which ontologies are analysed study the structural and typological aspects of the ontology graph. Here, "usage" is again considered from the standpoint of evaluating how the concepts are hierarchically arranged in the

ontology graph, but no insight into the "usage" of those ontologies is gleaned by creating relationships with ontology users.

Considering the above-mentioned observations, the main shortcomings of the existing approaches in the literature pertaining to measuring ontology usage are identified as follows.

1. Lack of a definition to describe ontology usage analysis.
2. Ontology usage has not been positioned as an area in the ontology engineering lifecycle.
3. No methodological approach for ontology usage analysis has been proposed.
4. Lack of a model to conceptually represent ontology usage analysis and make it accessible to others so that its analysis can be considered in the different areas of ontology engineering.

With these points in mind, in the next section we close the loop in the Ontology Lifecycle by positioning Ontology Usage Analysis in it and differentiating it from Ontology Evaluation and Evolution according to various criteria.

2.6   Positioning Ontology Usage Analysis in Ontology Lifecycle

It should be understood by now that analysing ontology usage on the Web is different from assessing and evaluating the quality of an ontology. Most of the means by which ontologies are modified, accessed or assessed are considered to be auxiliary rather than integral components of ontology development methodologies. The reason is because the research community working on ontologies is historically rooted in the knowledge engineering community, and the emphasis has therefore been on envisaging a conceptual representation of domain knowledge. Thus, most of the early work published under the rubric of ontology engineering focuses on the development (design-time) stage of ontologies (see Fig. 2.8) and little emphasis is placed on the in-use (run-time) stage of ontologies. By contrast, ontology usage analysis is concerned about the in-use stage, in which an ontology is viewed as a digital engineering artefact and its adoption, update and utilization is fully assessed. In the following sub-section, the aspects in which Ontology Usage Analysis (OUA) is different from its adjacent areas such as ontology evaluation, ontology maintenance and ontology evolution are discussed.

2.6.1 Ontology Usage Analysis (OUA) Versus Ontology Evaluation

Ontology Usage Analysis is different in many ways from Ontology Evaluation in spite of there being an overlap. The basic definition of OUA is that it analyses the use of ontology on the Web in a real world setting. Even though no formal definition for ontology evaluation is available in the literature, it is commonly referred to as a set of tools and methods to compare, validate and rank similar ontologies [181, 280]. Ontology evaluation and other ontology quality approaches [258] are important, however their principal emphasis is on guaranteeing that what is built will meet requirements (ontology developer's view) and that the final product (ontology artefact) is error free. Therefore, in some ontology engineering methodologies, ontology evaluation is a built-in process, while in others, it is considered as an independent component [40].

It can be seen that Ontology Evaluation focuses on the post-development phase of an ontology whereas OUA focuses on a post-implementation assessment scenario where the practical utilization of a particular ontology in the Semantic Web context is observed and its adoption, co-use and reuse is analysed after it has been instantiated. OUA focuses on the instantiated structured data on the Web, based on a domain ontology. For this reason, OUA can be viewed as a separate and independent activity from Ontology Evaluation. Ontology Evaluation helps in answering such questions as whether the built ontology matches the purpose, whereas Ontology Usage analysis provides the information needed to answer questions such as *"Given multiple choices, which ontology should I use to describe the (domain-specific) information on the Web?"* Ontology usage can help to identify the number of ontologies presently adopted and being used by different publishers, and their frequency of use assists in quantifying ontologies in terms of their usage. Therefore, OUA is a post-implementation process and a part of ontology maintenance which can assist in ontology evaluation, as explained in Sect. 1.5. In Table 2.1, OUA and Ontology Evaluation are compared with relevant factors to highlight the particular role and scope each has on ontologies.

2.6.2 Ontology Usage Analysis Versus Ontology Evolution

As mentioned earlier, the emphasis of OUA is to understand and measure ontology (vocabulary) usage in terms of its population, semantic relationship between different concepts, conformance with linked data principles and possible inferencing, depending on the axioms of the ontology. On the other hand, Ontology Evolution, which is closely related to ontology change and versioning, covers the change management process to keep the ontology artefact up-to-date and increase its effectiveness and usefulness. Ontology Evolution is defined as the timely adaptation of an ontology to changes which arise and the consistent management of these changes [132]. The

Table 2.1 Drawing comparison between ontology usage analysis, ontology evaluation and ontology evolution

	Ontology usage analysis	Ontology evaluation	Ontology evolution
Scope	Analyse how ontologies are being used	Evaluate how fit an ontology is to serve its purpose	Timely adaptation of an ontology to changes that arise [132]
Ontology lifecycle	Run-time	Design-time	Run-time (part of maintenance process in ontology development methodology)
Perspective covered	Usage, Structural, Semanticity, Incentives	Functional, structural, logical consistency, annotation property usage	Logical consistency, backward compatibility
Provide answers to	How to decide which ontology or term is suitable for reusability [295]?	How to measure the quality of an ontology for the Web [280] ?	How to evolve/update the conceptual model of an ontology?

sources of change that trigger ontology evolution are explicit requirements or the result of some automatic change discovery method. A comparison between OUA and Ontology Evolution, considering different factors, is presented in Table 2.1. In this regard, while both OUA and Ontology Evolution focus on the run-time phase of an ontology lifecycle model, they differ in scope. The current approaches [244] ignore an important source of information, that is, information about the actual utilization of an ontology on the Web. This utilization needs to be analysed, based on metrics and measurements, to qualitatively and quantitatively describe usage.

Even though the scope of each is different, Ontology Evolution and OUA can benefit from each other. For example, ontology usage analysis can provide the experiential evidence to gauge the anticipated impact of proposed changes in ontology. Recently, GoodRelations [142] has gone undergone several revisions[7] to evolve their conceptual model and implement changes in their model. Usage-based analysis provides the practical perspectives on ontology evolution which are obtainable through ontology usage analysis and which help to maintain logical consistency in an ontology.

2.7 Recapitulation

In this chapter, a survey of the existing literature relevant to the work of ontology usage analysis was presented. Two streams of work were evaluated: the first stream covered the work in which ontology is developed, and in the second stream cov-

---

[7]http://www.heppnetz.de/ontologies/goodrelations/v1.html; retr. 13/12/2017.

ered usage from the RDF (Semantic Web) data perspective. The relevant literature in each category was discussed to provide the necessary background and context to support identification of the gaps with respect to ontology usage analysis. The missing elements with respect to Ontology Usage Analysis in the Ontology lifecycle were discussed before the loop in the Ontology Lifecycle was closed by positioning Ontology Usage Analysis as one of its component parts. In the next chapter, Ontology Usage Analysis is formally defined, along with the Ontology Usage Analysis Framework (OUSAF), to ascertain the usage of different ontologies.

Chapter 3
Ontology Usage Analysis Framework (OUSAF)

3.1 Introduction

As emphasized in the previous chapters, web ontologies are being developed and deployed to describe information in a semantically rich fashion, but to benefit from the deployment of ontologies, it is important to understand which components of an ontology are being used and how they are being used. Such understanding can improve the utilization of Semantic Web data and allow its potential benefits to be realized [13]. In the previous chapter, we discussed the existing literature, which measures "usage" with a greater focus on understanding RDF data in general than from the perspective of analysing how ontologies are being used to represent that RDF data. To address that drawback, we present our proposed **O**ntology **US**age **A**nalysis **F**ramework (OUSAF) in this chapter.[1] Section 3.2 presents the notations used in this book. In Sect. 3.3, 'Ontology Usage Analysis' is defined and key terms used in the definition are discussed in detail. Section 3.4 describes the different phases of Ontology Usage Analysis framework and the purpose of each phase. In Sect. 3.5,

---

[1] Parts of this paper has been republished with permission from:

1. Jamshaid Ashraf, Omar Khadeer Hussain, Farookh Khadeer Hussain, A Framework for Measuring Ontology Usage on the Web, The Computer Journal, 2013, Volume 56, Issue 9, pp. 1083–1101, by permission of OUP.

2. John Wiley & Sons from Making sense from Big RDF Data: OUSAF for measuring ontology usage, Jamshaid Ashraf, Omar Khadeer Hussain, Farookh Khadeer Hussain, Volume 45, Issue 8, Copyright © 2014 John Wiley & Sons, Ltd; permission conveyed through Copyright clearance centre.

3. Knowledge-Based Systems Volume 80, Jamshaid Ashraf, Elizabeth Chang, Omar Khadeer Hussain, Farookh Khadeer Hussain, Ontology usage analysis in the ontology lifecycle: A state-of-the-art review, pp. 34–47, 2017.

4. John Wiley & Sons from Empirical analysis of domain ontology usage on the Web: eCommerce domain in focus, Jamshaid Ashraf, Omar Khadeer Hussain, Farookh Khadeer Hussain, Volume 26, Issue 5, Copyright © 2013 John Wiley & Sons, Ltd; permission conveyed through Copyright clearance centre.

© Springer International Publishing AG 2018

J. Ashraf et al., *Measuring and Analysing the Use of Ontologies*, Studies in Computational Intelligence 767, https://doi.org/10.1007/978-3-319-75681-3_3

each phase of the framework is explained in detail. The conclusion of the chapter is presented in Sect. 3.6.

3.2 Notations

In this section, the core notations used in the book are explained. The models of ontology and knowledge base used in this book are primarily based on [185]. For a more detailed background and formal discussion on RDF-related terms reader should refer to the following resources [139].

3.2.1 URI Reference

On the Semantic Web, all information has to be expressed as statements about Resources. A resource is identified by a Uniform Resource Identifier (URI). URIs identify not only Web documents, but also real-world objects like people and cars, and even abstract ideas and non-existent things like mythical concepts. All these real-world objects or things, are called resources in the Semantic Web, and the URI Reference is a compact string of characters for identifying an abstract or physical resource.

3.2.2 RDF Term

A RDF term is denoted as $RDFTerm := U \cup B \cup L$, where U is a set of URI references, B is a set of blank nodes and L is a set of literals. Furthermore,

- The set of **blank nodes** B is a set of existentially qualified variables.
- The set of **literals** is given as $L = L_p \cup L_t$, where L_p is the set of **plain literals** and L_t is the set of **typed literals**. A typed literal is the pair $l = (s, t)$, where s in the lexical form of the literal and $t \in U$ is a datatype URI.

The above-mentioned sets U, B, L_p, L_t are pairwise disjoint.

3.2.3 RDF Triple

A RDF triple t is defined as $t := (s, p, o)$ where s is the subject, p the predicate and o the object.

3.2.4 Data Level Position

Data level position identifies the position where instance data can be found on an RDF Triple. It refers to the position of the subject of a triple and object of the triple *iff* the predicate is not rdf:type.

3.2.5 Schema Level Position

Schema level position identifies the position where schema elements (terminological knowledge) can be found in an RDF Graph. It refers to the object *iff* the predicate in rdf:type and predicates other than rdf:type.

3.2.6 Ontology Structure (O)

An ontology structure is a 6-tuple $O := \{C, P, A, H_C, prop, att\}$ consisting of two disjoint sets C and P whose elements are called concepts and relation identifiers, respectively, a concept hierarchy $H_C : H_C$ is a directed, transitive relation $H_C \subseteq C \times C$ which is also called concept taxonomy. $H_c(C_1, C_2)$ means that C_1 is a sub-concept of C_2, a function $prop : P \rightarrow C \times C$, that relates concepts non-taxonomically (the function $dom : P \rightarrow C$ with $dom(P) := \Pi_1(rel(P))$ gives the domain of P, and $range : P \rightarrow C$ with $range(P) := \Pi_2(rel(P))$ gives its range. For $prop(P) = (C_1, C_2)$ one may also write $P(C_1, C_2)$). A specific kind of relations are attributes A. The function $att : A \rightarrow C$ relates concepts with literal values (this means $range(A) := STRING$).

3.2.7 Dataset (Ontology-Based Metadata)

A metadata structure is a 6-tuple $Dataset := \{O, I, L, inst, instr, instl\}$, that consists of an ontology O, a set I whose elements are called instance identifiers (correspondingly C, P and I are disjoint), a set of literal values L, a function $C \rightarrow 2^I$ called concept instantiation (For $inst(C) = I$ one may also write $C(I)$), and a function $inst : P \rightarrow 2^{I \times I}$ called relation instantiation (For $instr(P) = \{I_1, I_2\}$ one may also write $P(I_1, I_2)$). The attribute instantiation is described via the function $instl : P \rightarrow 2^{I \times L}$ relates instances with literal values.

3.2.8 Data Source

By data source we mean the unique pay-level domains (PLD) hosting RDF in some form (RDF document, RDFa snippets within HTML pages). The terms "data source", "site", "data provider", "data publisher" may be used interchangeably to refer to the "data source" that hosts RDF (in any serialization and form, i.e. RDFa) data.

In the next section, Ontology Usage Analysis is defined.

3.3 Defining Ontology Usage Analysis

As shown in Fig. 1.9, ontologies mainly go through two stages in their lifecycle. The first stage is the development stage which covers the set of activities relevant to the construction of the ontologies and their evaluation. The second stage is the in-use stage which covers the life span in which ontologies are being used to perform the intended tasks. This latter stage represents the run-time environment for ontologies which is described as Ontology Usage. Ontology Usage Analysis (OUA) is a task in this stage that provides insight into how ontologies are being used. This will lead to the better utilization of ontologies and effectual access to their instantiated data. Ontology Usage Analysis is defined as:

Definition (**Ontology Usage Analysis**), A study that examines the use of an ontology on the Web after it has been *instantiated* in a real world setting by measuring its *usefulness*, *usage* and *the commercial advantages* it offers.

This definition is comprised of many important terms (underlined) in measuring the usage of ontologies on the Web. Each will be explained in detail.

Instantiation

Instantiation means that a *term* defined in an ontology is being *used* in different usage scenarios (e.g. semantic annotation, knowledge representation, Semantic Web applications). The term could be a concept, or an object property (relationship), or a datatype property (attribute). Also, the term used refers to the event when an instance of a concept type is created, or when an object property is used to relate two individuals, or when a datatype property is used to associate data values. The instantiation of ontologies provides access to a corpus of semantically rich structured data comprising schema-level and instance-level data. Since the intrinsic value of ontologies is associated with their increased reusability [141], the instantiation of ontologies helps in attaining the increased perceived value and utility of ontologies in use. This provides the usage trends of ontologies to promote reusability. The (re)usability (being the utmost quality of any reusable component) of ontologies is facilitated by gaining an insight into how an ontology is actually being instantiated and used.

Usefulness

Usefulness means measuring the structural characteristics of ontological components to understand the distribution of relationships between different concepts and the attributive characteristics of the data properties. Measuring usefulness quantifies how the ontological model is conceptualized and organized structurally to arrange the relationship, including the taxonomical and non-taxonomical relationships with other concepts.

Usage

Usage refers to the state when an ontology is in use and it measures the statistical characteristics of the ontological components that are being used through ontology instantiation. The usage of an ontology helps in understanding how correctly the model is conceptualized to represent the real world entities and the relationship those entities hold. Usage encompasses the use of concepts, the use of object properties to create typed relationships with other entities and the use of certain attributes to describe entities.

Commercial Advantage

The *commercial advantage* quantifies the incentives available to the users of the ontology as a result of using it. This helps in incorporating the driving factors behind the adoption of the ontology by the users to further promote and encourage its reusability. It captures the benefits available to the adopters of ontologies in publishing semantically structured data on the Web.

In other words, ontology usage analysis provides qualitative and quantitative insight into how an ontology is being adopted, the common patterns of its usage in the real world setting, how useful it is and what benefits it provides. Based on the definition of OUA, we propose the methodological approach for analysing ontology usage in the next section.

3.4 Methodological Approach for OUA: Ontology Usage Analysis Framework (OUSAF)

The aim of this book is to develop and implement a framework for Ontology Usage Analysis framework known as **O**ntology **US**age **A**nalysis **F**ramework (OUSAF). To develop such a framework, there is a need to create a process that is complete and contains the necessary detailed descriptions for ontology usage analysis. To make the methodology more practical (that is, easy to follow in a real world setting), it should provide fine grained descriptions of the steps involved, the methods or techniques applicable, should assign roles to activities and present a clear idea about the input and outputs of the processes involved [235]. The developed methodological approach should provide the necessary detail to make it implementable; however, at the same time, it should be kept generic enough to allow the provision of different methods

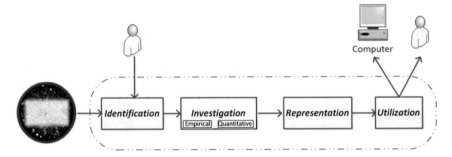

Fig. 3.1 Phases in ontology usage analysis framework

and techniques to be adopted for different application scenarios, when such a need arises. Keeping these requirements in view, we propose the OUSAF which has four broad phases, namely; *identification, investigation, representation* and *utilization* as shown in Fig. 3.1. Each of these phases is discussed in the forthcoming subsections.

3.4.1 Identification Phase

Identification phase refers to the selection of the ontologies to be analysed. Identification of ontologies for usage analysis can be for; *(a)* to determine the usage analysis of a specific domain ontology already known in the application area, for example FOAF for social networking; and/or *(b)* to investigate and identify the interesting ontologies available in the domain-specific dataset. These two scenarios require different solutions for identification. The former type of scenario is trivial and can be addressed by looking up the available semantic search engines to directly identify the namespace of the domain ontology. However, the latter scenario needs an exploration mechanism to identify how different ontologies are being used in a dataset, or those which are pertinent to domain-specific application areas. Therefore, before proceeding with the usage analysis, the candidate ontologies need to be identified from the corpus of the dataset. The aim of this phase of OUSAF is to develop the technique and approach to identify the potential candidate ontologies for usage analysis. Further discussion on this phase is in Sect. 3.5.1.

3.4.2 Investigation Phase

Investigation phase refers to the analysis of the use of a particular ontology. The aim of this step is to analyse the identified ontology to measure its usage and the patterns of its usage. The usage analysis investigates how the conceptual model and the ontology components, such as Classes, Relationships, Attributes and Axioms, are being used to annotate real world data. To obtain a comprehensive insight into the usage of a given ontology, the analysis is required to be performed at two levels. In the first level, empirical analysis needs to be performed to understand and highlight the key aspects that contribute to the usage of ontologies. In the second level, based on the key aspects obtained through empirical analysis, ontologies need to be quantitatively analysed. The dimensions related to the usage of ontologies considered in the investigation phase: are (a) usefulness; (b) usage; and (c) commercial advantage. The obtained results are then combined to ascertain the usage analysis. From a methodological point of view, it is important to point out that the key requirement of the framework is to support the different techniques and methods required to measure the statistical properties of ontology adoption. This requirement allows for the adoption of feasible support methods, tools and techniques to improve the applicability and effectiveness of the usage analysis in a real world setting. Further discussion on this phase along with the metrics used in this phase are given in Sect. 3.5.2.

3.4.3 Representation Phase

The purpose of investigating ontology usage is to understand how an ontology is being used by different users and to exploit this information to utilize Semantic Web data effectively and efficiently. Therefore, analysis results obtained in the investigation phase have to be represented in a structured format to allow a larger number of applications to use it for further information processing. This is done in the *representation* phase, in which analysis results are represented for further exploitation. Information processing, in this context, may include information retrieval, interlinking with other datasets, mash-ups and the automatic generation of prototypical queries. Additionally, for the optimal utilization of analysis in Semantic Web, the results need to be represented in a format which is equally accessible to both human and machine actors. Further discussion on the representation phase is Sect. 3.5.3.

3.4.4 Utilization Phase

Utilization phase refers to that phase in which the output of the usage analysis is further utilized to achieve conceivable benefits. Since there are different scenarios in which ontology usage analysis is helpful (as discussed in Sect. 1.7), the utilization phase covers the activities related to the exploitation of results, by different use case scenarios. To facilitate the utilization of the analysis in different application areas, the results are represented through a structured format developed in the representation phase, allowing the wider dissemination and exploitation of findings. To improve the usability of the methodology, use cases are implemented which use the ontology usage analysis information to assist applications in either accessing precise information from the Web or the assimilation of information to offer wider perspectives. Further discussion on this phase is in Sect. 3.5.4.

In the next section, various steps that need to be achieved in each phase of OUSAF are discussed.

3.5 Details of Each Phase of OUSAF

The main role of the OUSAF (depicted in Fig. 3.2) is to empirically analyse the RDF data on the Web with a focus on domain vocabularies and ontologies. The framework supports conducting empirical analysis from two dimensions: one from an ontology perspective and the second from the RDF data perspective. From the ontology perspective, we consider the ontology as an engineering artefact (ontology document) to characterize the components defined in the document such as vocabulary, hierarchal and non-hierarchal structure, axioms and attributes. From the RDF data perspective, RDF triples are analysed to understand the patterns and structure of the data available in the dataset. As mentioned in Sect. 3.4, the methodological approach followed for the analysis comprises four different phases, namely: *identification*, *investigation*, *representation* and *utilization*. In Fig. 3.2, which provides the schematic diagram, each of the steps is marked using a dotted rectangular box. The detail for each phase is discussed in the following subsections.

3.5.1 Identification Phase: Identification of Candidate Ontologies

As mentioned in Sect. 3.4.1 earlier, there are two different ways by which ontologies for usage analysis are identified; first, the domain ontology which needs to be evaluated is known or given; and second, ontologies being used in the corpora/dataset need to be identified.

3.5.1.1 Usage Analysis of a Specific Domain Ontology

In this scenario, a user would like to analyse the domain ontology which conceptually represents the application area of interest. There are two possibilities: first, the user knows the specifics of the required domain ontologies such as the namespace of the ontology and the URI hosting the formal authoritative document of the ontology; and second, the user would like to search for a specific domain ontology. The first case is trivial and for the second case, the user can search for the domain-specific ontology using different services such as ontology search engines (Swoogle [79], Watson [68]), ontology libraries (OntoSelect [50], Cupboard [67], BioPortal [206]),

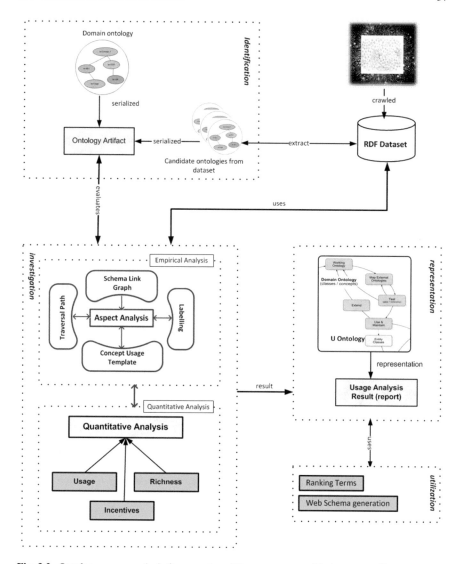

Fig. 3.2 Ontology usage analysis framework and its components with the process flow

Semantic Web search engines (Sindice [266], SWSE [149]), and other applications built on Linked data corpora (FactForge [28], Sig.ma [265]). Almost all of these services return the URI of the ontology to retrieve the authentic ontology source document.

3.5.1.2 Identify and Analyse Candidate Ontologies from Dataset:

It is also practically desirable to investigate the prevalence of different ontologies in a vertical application domain. This helps to know what different but related ontologies are being used to conceptualize the domain data. It is more advantageous from the viewpoint of data publishers to know the availability of different ontologies being used by the community to describe the information on the Web. To address such common requirements, we propose the use of a domain-specific dataset to be used for the identification of ontologies and their usage in the dataset. The use of a domain-specific dataset will not only help in identifying the ontologies, but will help to obtain the level of consensus that exists in the use of these ontologies. However, it is important to note that in such an approach, the considered dataset needs to be representative of the actual RDF data available on the Web and this aspect should be considered while interpreting the identification of ontologies and their usage in general.

To obtain such insight, **Ontology Usage Network (OUN)** is constructed to model the use of ontologies by different data publishers. The ontology usage network is based on the **Affiliation Network** model [38] which captures the affiliation of agents to societies. In other words, it provides an intuitive way of representing participation data (membership) and studies the dual perspectives of the actors (ontologies, in our case) and the events (data publisher). Using Social Network Analysis (SNA) techniques, the OUN is processed and analysed to obtain the following insight:

- understand the use of different ontologies by different data publishers
- understand the ontology-to-ontology relationship present based on the co-participation of ontologies in different data sources

In order to obtain such insight, the OUN needs to be processed to generate a representational model capable of providing the required information. Undertaking such an analysis will assist in obtaining a better understanding of current ontology usage patterns and the similarities in usage among different data publishers. In Chap. 4, the Ontology Usage Network Analysis Framework (OUN-AF) is presented to facilitate the identification of the ontologies from the dataset and perform the analysis to obtain the required insight. The framework implements different metrics and techniques and methods to structurally, typologically and semantically analyse the OUN and identify different ontologies and their co-participation behaviour. The methods and techniques followed for identification are dataset agnostic, thus making the approach applicable to different domain-specific datasets.

3.5.2 Investigation Phase: Investigating the Ontology Usage

To measure ontology usage, the dataset which comprises the semantic data collected from the web-of-data is considered and the instantiation and the use of properties of the conceptualized domain are measured, modelled by the domain ontology. As mentioned in Sect. 3.4.2, ontology usage is investigated at two levels: first empirically and then quantitatively. For these two levels of investigations, two frameworks are developed as part of OUSAF. These two frameworks are **EMP**irical Usage **A**nalysis Framework (**EMP-AF**) and **QUA**ntitative Usage **A**nalysis Framework (**QUA-AF**). They are briefly discussed next:

3.5.2.1 Empirical Analysis

The use of different ontologies is empirically analysed to understand the key aspects involved in ontology usage. The EMP-AF utilizes a set of metrics to measure the use of different ontology components from different aspects to establish a better insight into the prevailing usage patterns on the Web. The different aspects considered are the use of pivot concepts, their semantic description, the use of textual description and knowledge and data patterns. The metrics developed for these aspects are as follows:

- **Schema Link Graph**: Schema Link Graph (SLG) unveils the relationships that are present between different vocabularies at instance level to semantically describe the entities being represented by the ontology concepts.
- **Concept Usage Template**: Concept Usage Template (CUT) captures the instantiation of concepts, the relationship it has, and the use of different data properties to provide factual information.
- **Labeling**: Labelling captures the use of different properties for labelling purposes. Labelling properties helps by providing a textual description of the entities useful for human interpretation and user interface.
- **Traversal Path**: Traversal path captures the data and knowledge patterns prevalent in the dataset.

3.5.2.2 Quantitative Analysis

Based on the insight obtained from the empirical analysis, ontologies in quantitative analysis are analysed from different dimensions to obtain a comprehensive insight into their usage. The QUA-AF implements metrics that are grouped into three categories to measure the usage of an ontology, encompassing its *usefulness*, *usage* and *commercial* advantages. Usefulness measures the richness of the ontology components and provides structural insight into how a given ontology is modelled and how the semantics are represented. While on one hand, the inclusion of such information helps in identifying the semantically rich components of an ontology, on the other hand, it also assists in drawing a comparison, if any, between the usage and semantically rich components. Usage mainly captures usage patterns in terms of the presence of different ontological terms in describing the instance data to provide semantic metadata on the Web. Commercial advantage captures the incentive model available to early ontology adopters and Semantic Web data publishers. It considers all the components of the ontologies such as classes, relationships, taxonomical relationships and axiomatic triples to quantify usage trends on the Web.

- **Measuring Ontology Richness**: In this category, the richness of the ontology components such as concepts, object properties (relationship) and data properties (attributes) are quantified. In the case of RDFS vocabularies, since object and data properties are not disjoint, only object properties are considered to refer to the predicates defined by the vocabulary.
- **Measuring Ontology Usage**: In this category, use of ontologies and their components which includes the use of different concepts, relationships, attributes and axioms is measured.
- **Measuring Incentive**: In this category, the key factors fostering the growth and adoption of vocabularies/ontologies are considered as the driving factors for early adopters are measured. Two of the other driving factors to consider in this research are the incentives available to structured data publishers and the support available for an ontology/vocabulary in Semantic Web applications and tools.

These sets of metrics provide a more practical view of the use of ontologies since they cover the technical aspects of the ontology (usefulness), adoption and uptake of the ontology (usage) and the incentives available for ontology users (commercial advantages), thereby covering all aspects which, if considered, can help to identify compelling products [234] which, in our case, are ontologies.

- **Combining the Analysis Results**: The quantified measures of the above-mentioned aspects are combined to obtain an overall picture of the usage of a given ontology. Combining these values further helps in ranking them to obtain the required set of ontology components, based on user requirements.

The metrics developed for the different stages of investigation phase are discussed in Chaps. 5 and 6. The developed metrics cover the different aspects of usage to obtain detailed insight into the required quantitative measures which are useful for the exploitation of the results. Similar to identification, the methods and techniques used for investigation are ontology and dataset agnostic, therefore making the solution applicable to different ontologies and datasets.

3.5.3 Representation Phase: Representing Usage Analysis

The representation phase of the usage analysis methodology concerns the representation of results in such a way that they can be easily disseminated and accessed by other applications. To capture the analysis results, Ontology Usage Ontology (U Ontology) is developed. U Ontology is a meta-ontology which provides the conceptual architecture to represent the usage patterns of the domain ontology in a dataset. The usage patterns contain both the knowledge and data patterns which assist in under-standing the knowledge available in the dataset and generate prototypical queries to access data. In other words, U Ontology provides machine processable information which can be used to improve the accessibility of Semantic Web data and the reuse of ontologies. The usage analysis of a particular domain ontology obtained using OUSAF is encoded using U Ontology which provides a different set of concepts, and relationships between concepts, to allow the user to access the analysis find-ings programmatically. The availability of information on how ontologies are being used and what the prominent knowledge and data patterns are helps in effectively accessing the required information over the Web. Additionally, such metadata pro-vides the meta-level information about ontologies, including their usage, to support application/tool development and provide pragmatic feedback to ontology evolution and change management.

Following the Semantic Web community recommendations of reusing existing ontologies wherever possible, the U Ontology is considered as an extension to the OMV (Ontology Metadata Vocabulary). OMV [137] attempts to provide a standard ontology metadata for describing ontologies and their entities. The metadata vocab-ulary for describing ontologies is modelled as an ontology and is called OMV Core[2] with the provision of supporting different extensions [216]. The U Ontology is con-sidered one of its extensions, implementing usage analysis of ontologies to further enrich existing application-specific ontology-related information such as mapping, ontology evaluation, and ontology changes. In Chap. 7, the conceptual model devel-oped to provide the representation for usage analysis is presented. A formal concep-tualization model, based on RDF and OWL that allows the standardized formulation of ontology usage analysis results, is adopted.

3.5.4 Utilization Phase: Utilizing Usage Analysis Results

The *utilization* phase makes use of the analysis results. As mentioned in Sect. 1.7, usage analysis can be used by different groups of users (ontology developers, data publishers and data consumers) to access information over the Web by generating prototypical queries based on the schema-level data available in the dataset. The U Ontology, which is populated during the representation phase, contains usage-related information for access by users to know about the usage of an ontology.

The populated U Ontology provides descriptive and quantitative details on the use of different ontology components which is useful for realizing the benefits of ontology usage analysis. Two use cases, namely: (a) construction of prototypical queries; and (b) construction of Web Schema, are explored in the utilization phase to demonstrate the utility of usage analysis. In the case of prototypical query gen-eration, U Ontology provides the list of ontology components with their usage to assist in querying explicit and implicit information. Similarly, in the case of Web

---

[2]http://ontoware.org/projects/omv; retr., 19/10/2017.

Schema, highly used vocabularies for a given vertical application area are accessed to understand the structure of the domain-specific entities. Further explanation on how results from the usage analysis are obtained are mentioned in Chaps. 5 and 6. The formal model with instance data (ontology usage analysis results) is then used to present the ontology usage catalogue, encapsulating the usage status of a given ontology, which is discussed in Chap. 8.

3.6   Recapitulation

In this chapter, the OUSAF and its components were discussed. Details of the methodological approach were presented to specify the phases involved in carrying out the empirical analysis: *identification*, *evaluation*, *representation* and *utilization*. In the next chapter, the model used to construct the Ontology Usage Network is discussed, as well as Social Network Analysis techniques and methods used for the identification phase of the OUSAF.

Chapter 4
Identification Phase: Ontology Usage Network Analysis Framework (OUN-AF)

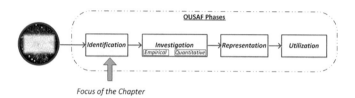

Focus of the Chapter

4.1 Introduction

Analysing ontology usage, as we have noted in the previous chapter, is comprised of four phases: *identification, investigation, representation* and *utilization*. The identification phase, which is the focus of this chapter,[1] is responsible for identifying different ontologies that are being used in a particular application area or in a given dataset for further analysis. The ontologies to be analysed fall into two categories:

- the domain ontology to be analysed for usage is known
- the domain ontology to be analysed for usage needs to be identified according to application-specific requirements.

The first category is trivial because the user can access the ontology from its respective namespace URI; however, in the latter case, a mechanism is required to identify the presence of different ontologies in the required domain and to select the potential ontologies based on the users specific requirements and selection criteria.

---

[1]Parts of this chapter have been republished with permission from:

© [2017] IEEE. Analysing the Use of Ontologies Based on Usage Network, 2012 IEEE/WIC/ACM International Conferences on Web Intelligence and Intelligent Agent Technology, 4–7 Dec. 2012, Macau, pp. 540–544.

LNCS on Web Technologies and Applications, Ontology Usage Network Analysis Framework, volume 7808, 2013, pp 19–30, Jamshaid Ashraf and Omar Hussain, Springer-Verlag Berlin Heidelberg 2013 "With permission from Springer".

© Springer International Publishing AG 2018
J. Ashraf et al., *Measuring and Analysing the Use of Ontologies*, Studies
in Computational Intelligence 767, https://doi.org/10.1007/978-3-319-75681-3_4

Some of the common requirements which form the selection criteria for the identification of ontologies in this scenario can be established in response to the following
questions:

1. Which are the widely used ontologies in the given application?
2. Which ontologies are more interlinked with other ontologies to describe domain-
    specific entities?
3. Which ontologies are used more frequently and what is their usage percentage
    based on the given dataset?
4. Which ontology clusters form cohesive groups?

    To analyse these selection criteria and, to identify different ontologies, their links
with other ontologies, the usage patterns prevalent in an application-specific area, and
detailed insight into which data sources (data publishers) use particular ontologies
is required. To establish a better understanding of ontology usage and to identify
the ontologies, based on the above-mentioned criteria, this chapter proposes the
Ontology Usage Network Analysis Framework (OUN-AF) which models the use of
ontologies by different data sources using an Ontology Usage Network (OUN). OUN
represents the relationships between ontologies and data sources based on the actual
usage data available on the Web in the form of a graph-based relationship structure.
This structure is analysed using metrics to study the structural, typological and functional characteristics of OUN by applying Social Network Analysis (SNA) [167]
techniques.

    This chapter is organized as follows. Section 4.2 introduces Social Network
Analysis (SNA) and the different types of relationships often represented in SNA.
Section 4.3 provides the rationale for using SNA to obtain the required analysis for
the ontology identification phase. It also provides an overview of the literature in
which SNA has been used in the context of ontologies. In order to provide the background and introduce terms relevant to SNA, Sect. 4.4 presents the key terms related
to SNA and the different types of networks and properties observed in these networks. In Sect. 4.5, the Affiliation Network, its graphical representation and relevant
concepts are detailed. In Sect. 4.6, OUN-AF is proposed for the ontology identification phase of OUSAF. OUN-AF phases and their sequence are presented, together
with a set of activities. In Sect. 4.7, the metrics developed to analyse the relationship
between ontologies and the data source are given. Section 4.8 gives an overview of
the analysis by applying the metrics on OUN and the projected networks. In Sect. 4.9,
the evaluation of the ontology identification phases based on OUN-AF is presented.
Lastly, Sect. 4.10 concludes the chapter.

4.2 Social Network Analysis

Social Network Analysis (SNA) is a methodical approach for mapping and measuring the relationships between people, organizations, computers, and information
resources. Historically, it belongs to the social sciences in which the social relationships among a set of actors were studied. However, in the past few years, the idea of

networks has been extended to include other unifying themes to study social inter-action in living species, digitally connected devices and natural world connections. As a result of this change, research in SNA is witnessing a substantial shift in its focus from small networks to large scale networks that are big in size and complex in structure. In general, SNA studies the social relationships between actors, and these relationships take different forms, depending on the type of network under study. More importantly, SNA provides the methods to characterize the structure of social networks, the important positions in the network, the strength of relation-ships between different sets of nodes and the existence of sub-networks [90]. In other words, SNA allows to measure the relationship, communication, and infor-mation flow between nodes through edges, and focuses on uncovering the patterns of actors' interactions in the network. Therefore, network analysis is based on the intuitive notion that these patterns are important features relating to the activities of the individuals who display them through their interaction [102].

Social networks are made up of actors that are linked by social *relationships*. Thus, actors and relational ties (links) are the basic elements of the network, in which a wide range of social relationships can take place between actors. The interlinking between nodes denotes the flow of information that reflects their social relationships. The different types of relationship have been studied in the literature [90, 111] and can be grouped into the following three main categories:

- Explicit relationships
- Interaction
- Affiliation

Each relationship category is briefly described in the next subsections.

4.2.1 Explicit Relationships

Explicit relationships represents the relationships between people, between organi-zations, or between people and organizations. For example, the explicit relationship between people could be one of brother, sister, parent, friend, etc. One of the earli-est research studies to examine explicit relationships was conducted by Zachary et al. [291]. In this study, a network was formed in a university-based karate club to understand the cause of internal conflict within the club (see Fig. 4.1). The explicit relationships between cohesive groups were identified as being due to social links and eventually, the club was split into two groups to mitigate the issues within the club. In other work, Hogg et al. [151] presented an empirical study of an online political forum where users engage in content creation, voting, and discussion, which forms the explicit connection in the network.

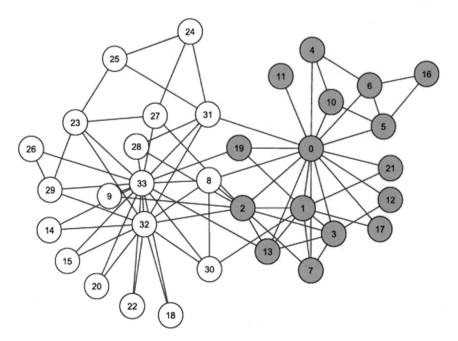

Fig. 4.1 Zachary club network. The links between two nodes represent explicit relationships and nodes are split into two groups, one of which is denoted by a white circle, the other by a grey circle

4.2.2 Interaction

Interaction represents the flow of information or communication between actors, such as the relationship between authors or the interaction of players on a team. In the digital world, this takes the form of participation and collaboration in discussion forums, and co-authoring articles in wikis. Brass et al. [41] investigated the interaction patterns of men and women in an organization and the relationships between these patterns to study the perceptions which influence promotion to supervisory positions in the organisation. Interaction networks have been studied in biomedicine to understand the interaction between different chemical components. One such interaction-based network [260] analyses the protein-to-protein interaction networks (see Fig. 4.2) derived from phage-display and two-hybrid analysis.

4.2.3 Affiliation

Affiliation represents the similarity between actors in the network. Similarity emerges from the fact that actors share the same attributes, which enable the affiliation between them. The network which represents affiliation-type relationships is termed the "affil-

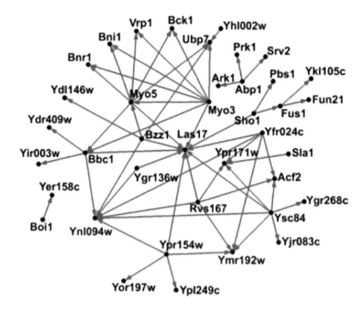

Fig. 4.2 Protein to Protein interaction network

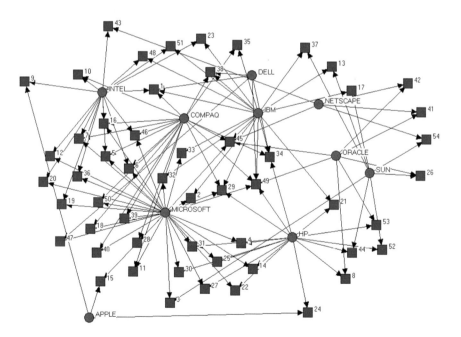

Fig. 4.3 Affiliation network (two-mode graph) consisting of 10 US computer and software firms (red circles) and 54 strategic alliances (blue squares)

iation network". Affiliation networks have been used extensively in economic science and social networks to study the affiliations between different but related elements, based on their affiliations. For example, in economic science, affiliation networks are used to analyse how organizations and their members interact to understand the mechanism of economics [187]. Wasserman et al. [284] presented an affiliation network (see Fig. 4.3) which was produced from the 1998 GIS dataset, consisting of 10 US computer and software firms and 54 strategic alliances between them. Affiliations networks are discussed in more detail in Sect. 4.5, due to their relevance to the model adopted for identifying the ontologies.

Using these relationships in a network, the underlying data source can be represented as a network-based format with relationships between different elements, thereby allowing analysis to be carried out on it. In the next section, the rationale for using SNA and its techniques to identify ontologies from a given dataset in the ontology identification phase is discussed.

4.3   Rationale of using SNA in Ontology Identification

As mentioned in Sect. 4.1, the aim of the ontology identification stage is to identify the potential ontologies on which a detailed usage analysis needs to be performed. Several criteria may be required to assist the selection of an ontology from a given dataset. Some of these are:

1. To select the ontology used by the highest number of data publishers in a given domain.
2. To determine what other ontologies are being co-used by the same data publishers.
3. Given the other co-used ontologies, to determine which ontology holds the central position among them. (This analysis is beneficial for data publishers to understand which ontologies play a central role in facilitating the use (or even adoption) of other ontologies.)
4. To establish whether there are any common ontology usage patterns among the data sources (data publishers) that dominate the considered dataset.

These selection criteria are applied on a dataset that is highly complex in nature, large in volume and highly interconnected. Figure 4.4 shows the latest version of the Linked Open Data (LOD) cloud currently published on the Web. From this figure, it can be seen that various relationships exist between the different datasets. It is important to note that the LOD cloud diagram (Fig. 4.4) is a high level depiction of the interlinking between the datasets and encapsulates the underlying mechanisms used to create relationships between different datasets. Ontologies that are present at a lower level (not depicted in the figure) facilitate the semantic representation of the information and have interlinked entities distributed across different datasets. Undoubtedly, a dataset that comprises data from different sources, annotated using domain ontologies, is a complex network structure, and to study such a complex network structure according to the identified selection criteria requires a model capable

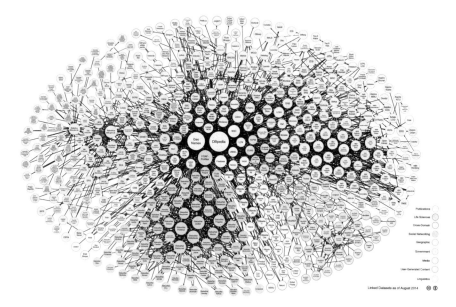

Fig. 4.4 Linking open data cloud diagram by Richard Cyganiak and Anja Jentzsch. http://lod-cloud.net/; retr. 12/11/2017

of representing multi- or bi-dimensional information components as a base model, in order to address the queries mentioned earlier.

Specific to this case, *an **ontology** can be used by any number of users, and a **user** can use any number of ontologies at the same time*. This introduces specific requirements that need to be considered in modelling the information, as follows:

1. Two types of entity are involved, namely 'ontologies' and 'users', and both of these entities can have connections with other types of entity. For example, a user (e.g. data publisher) can use a number of ontologies and similarly, an ontology can be utilized by several users. Therefore, the framework required to represent the interrelation between ontologies and users in the given dataset should be able to capture and represent multi-edge systems to allow an edge span over several nodes, in contrast to a normal graph where an edge connects only two nodes.

2. These two sets of entities are disjoint, because edges flow between these two sets of nodes rather than within their own set. This means that an ontology cannot have a relationship (direct connection) with another ontology, and a user (data publisher) cannot connect to another user in the network. (Note: Ontologies can import other ontologies to reuse the term; however here, the scope of the term 'use' refers to the use of ontologies by the data publisher for semantic annotation.)

Keeping in mind the above-mentioned complex network to be analysed in a dataset and the specific features according to how they should be represented, a framework is needed that is able to represent the information and relationships of a distinct set of entities within a dataset, in a format that can be extensively analysed. The framework should be able to identify ontologies and their use by different data sources, therefore the flow between different types of nodes needs to be studied to unleash the co-membership relationships which are otherwise not visible. One possible way to represent this kind of information in the required format is **Social Network Analysis (SNA)**. SNA provides the lens through which complex social networks and their components can be studied to mine the hidden relationships within their structure. The use of ontologies by different data sources resembles several social networks in which similarities between actors are frequently a source of interaction. We use Social Network Analysis (SNA) techniques to study the complex networks representing the information pertaining to the use of ontologies by different data sources.

In the next section, the terms and concepts related to the representation and analysis of Social Network Analysis are introduced prior to a discussion on the different types of networks.

4.4 Key Concepts of SNA Relevant to Ontology Identification Phase

Networks, which are also known as "graphs" in mathematical literature, represent the complex systems of interconnected components. In its simplest form, a network is comprised of *nodes* and *edges*. SNA provides mathematical techniques to quantitatively analyse the network and understand the relationship patterns within the network. In our approach, we use SNA methods and techniques to identify ontology usage patterns. The terms used in the discussion are described in the following subsection.

4.4.1 Key Terms and Their Definitions

Key terms used in our approach relevant to Social Network Analysis are as follows:

Network: a distinct set of actors and the connections between them

Graph: a distinct set of nodes and a set of edges

Node: basic unit of a network (or graph) which represents actors, also referred to as vertex in the literature.

Edges: a connection between two nodes. In social networks, edges are known as links representing relationships between two actors.

Hyperedge: an edge that connects more than two nodes

Weighted edge: an edge with as assigned value representing the importance of the edge

Labelled edge: an edge with a label attached to it to provide a description of the relationship

Hypergraph: a graph in which generalized edges (called hyperedges) may connect more than two nodes

Multigraph: the term multigraph refers to a graph in which multiple edges between nodes are permitted [135]

Weighted graph: when each branch of a graph is given a numerical weight, it is termed as a weighted graph [286].

Labelled graph: a graph in which each node is labelled differently (but arbitrarily), so that all nodes are considered distinct for the purpose of enumeration [286].

One-mode networks: one mode (1-mode) networks involve relationships between a single set of similar actors.

Two-mode networks: two mode (2-mode) networks involve relationships between two sets of nodes

Distance: represents the number of distinct ties (lines) required to reach between any two nodes according to the shortest route.

Path: a list of nodes of a graph, each linked to the next by an edge. Formally it is defined as: a path on a graph, also called a trail, is a sequence $\{x_1, x_2, ..., x_n\}$ such that $(x_1, x_2), (x_2, x_3), ..., (x_{n-1}, x_n)$ are graph edges of the graph and the x_i are distinct [286]

Direct Path: a sequence of directed edges from a source node to an end node. Formally, it is described as a sequence of vertices, v_1, v_2, v_n, in a directed graph such that there is an edge from v_i to v_{i+1} for $i = 1, 2, ..., n - 1$.

Geodesic or short path: the shortest sequence of edges between two given nodes.

Degree: degree of a node v represents the number of nodes it is immediately adjacent to. Mathematically, degree k_v of a node v is represented as:

$$k_i = \sum_{v=1}^{n} a_{ij}, \qquad 0 < k_i < n \tag{4.1}$$

where a_{ij} is the entry of the ith row and the jth column of the adjacency matrix A

k-core: A graph's k-core is a numerical value which represents the maximal subgraphs in which each node has a degree of at least k. The coreness of a node is k if it belongs to the k-core but not to the (k+1) core.

Density: density ρ, in general measures the connectedness in a network. Therefore, a high p value indicates a dense network and a low value indicates a sparse network. It is defined as:

$$\rho(G) = \frac{m(G)}{m_{max(G)}}, \qquad 0 < \rho < 1 \tag{4.2}$$

where m is the number quantifying the edges in the network and $m_{max(G)}$ quantifies the network's maximum possible edges. The maximum possible edges for an undirected and directed network are $\frac{n(n-1)}{2}$ and $n(n - 1)$ respectively.

Fig. 4.5 Power-law
distribution

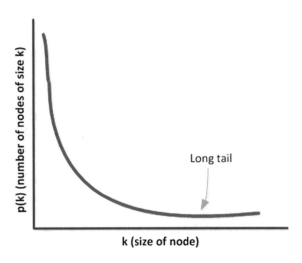

Centrality: a general measure of the position of an actor within the overall structure
of the social network. Centrality measures assist in answering the question, "who
is the most important or central actor (node) in the given network" [200]. There are
several metrics to measure centrality, the most widely used being degree centrality,
betweenness centrality and closeness centrality.

Power-Law Degree distribution: as mentioned earlier, a node's degree quantifies
the number of adjacent nodes to it. The probability distribution of network's nodes
degree values is termed as the degree distribution, $p(k)$. Real-world networks are
quite different from random networks in terms of degree distribution. Random
networks often show binomial degree distribution [202] because of the equal
probability of an edge being present or absent in the network. However, the degree
distribution in real-world networks is noted to be heterogeneous and highly right-
skewed [15]. This means that a large number of nodes have low degree and a
small number of nodes have high degree. In the literature, this is also known as
long tail, as shown in Fig. 4.5.

4.4.2 Types of Networks

A network, in its simplest form, is a set of nodes with edges between them. In the
literature, nodes are also referred to as vertices, and edges as ties and relationship
links. An example of a network which contains eight nodes and eight edges is shown
in Fig. 4.6. Networks are primarily composed of nodes joined by edges, but there
are other ways in which networks can be more complex in structure and topology.
Both nodes and edges can have a variety of properties which make the corresponding
networks more complex than the one shown in Fig. 4.6. An edge can have a direction

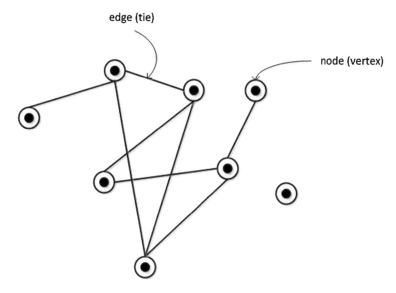

Fig. 4.6 A simple example of a network (with eight nodes and edges)

(which could be uni- or bi-directional), label, weight and attributes, and likewise, nodes can have type, weight and attributes as well. Other types of networks and their variations are discussed next.

4.4.2.1 Labelled and Directed Networks

A slightly more complex network is presented in Fig. 4.7. Figure 4.7a represents a graph with labelled nodes (i.e. A, B, C, D, E, F) and the edges are unidirected to show which two nodes are directly connected. In the context of social networks, the nodes can represent anything, such as a man, woman, boy, girl, city, or country; likewise, edges can show the kinship, friendship, professional affiliation, distance or other feature representing the relationship (tie) between nodes. In a network, either one or both nodes and edges can carry weights, which makes the network a weighted graph.

4.4.2.2 Labelled, Weighted and Bi-Directional Network

The additional attribute of weight can be added to the network. A weight can be attached to a node or an edge or both. Weight attributes can represent any quantifiable measurement necessary for the interpretation of the information represented in the network. For example, Fig. 4.7b represents a labelled, directed graph in which edges carry weight to represent the distance between cities.

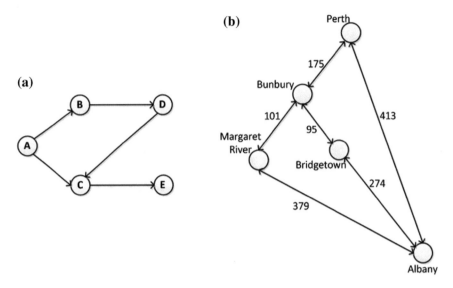

Fig. 4.7 Examples of different types of network: **a** network with labelled nodes and directed edges; **b** network with labelled nodes, weighted and bidirectional edges

4.4.2.3 Hypergraphs

Hypergraph is a special type of graph, in which edges join more than two nodes together. In the graphs of Figs. 4.6 and 4.7, the edges connect two adjacent nodes, whereas in a hypergraph, an edge (also known as a hyperedge) is incident to an unspecified number of nodes. Hyperedges are often used in social networks to indicate family ties. For example, all the individuals belonging to one family in a graph can be joined through a hyperedge which connects all the nodes representing individuals belonging to a family. In Fig. 4.8, a hypergraph with three hyperedges numbered 1, 2 and 3 is shown. Hyperedge 1 joins nodes B, D and E, hyperedge 2 joins E, F and G, and hyperedge 3 joins A, B and C. Hypergraphs are more expressive than regular graphs which often fall short in providing complex relational object representation [294]. Real world complex problems requiring the clustering and classification of objects based on their attributes are best represented as a hypergraph because of their expressivity, and to avoid information loss.

4.4.2.4 Bipartite (2-Mode) Graph

While hypergraphs help in clustering and partitioning nodes based on their attributes, there are other special graphs which are naturally partitioned in various ways. Such graphs are called **bipartite graphs** and contain nodes (vertices) of two distinct types, with edges only between nodes of a distinct type (See Fig. 4.9). **An affiliation network** is an examples of a bipartite graph in which actors and events are two types

Fig. 4.8 Hypergraph with three hyperedges

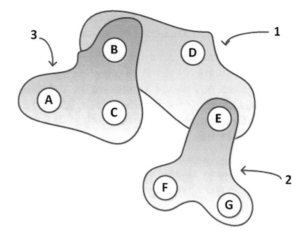

Fig. 4.9 An example of an author-paper affiliation network

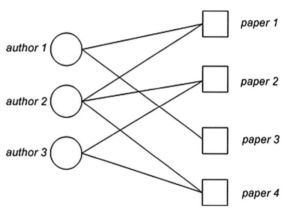

of entities (two sets of distinct type) related by the affiliation of the former in the latter [175].

The affiliation networks, their structure, key concepts, representational model and operations, such as 'projection', are presented in the next section.

4.5 Affiliation Network

A wide range of social networks has been built by analysing the similarities between actors, since these similarities are the source of the interaction between actors. This interaction, based on the similarity notion, is termed "*affiliation*" and a network representing such similarity-based relationships is called an affiliation network. The term "affiliation" is reserved for cases in which the data reflects some kind of participation or membership [38] in the social network. Affiliation networks are different in many ways from other types of network, as follows [284]:

- Affiliation networks are essentially two-mode (2-mode) networks
- Affiliation networks consist of subsets of actors, rather than a simple pair of actors
- The connection between members of the first mode (first set of nodes) is based on the relationship established through the second mode (second set of nodes)
- Affiliation networks provide the mechanism for studying the dual perspectives of actors and events

In an affiliation network, the importance of individual relationships within its society is studied to explore the behaviour of individuals and their acceptance by the society. The modelling of relationships using affiliation helps in identifying joint participation in social events and provides the opportunity for individuals to develop pairwise relationships with other individuals based on their participation. In the context of ontology usage, an affiliation network is used to model the use of an ontology by different data publishers. The ontologies are the actors and the data publishers form the hypothetical society (i.e. event) in which these actors participate. Affiliation networks cannot be analysed by merely looking at the pairs of the actors or events because it is subsets of actors who participate in the events. However, it is often desirable to understand the patterns of relationships within one of the sets. In other words, this mean to know how two nodes of the same set are related to each other, based on their relationship to the other set of nodes. These kinds of relationship, which are inferred based on the relationship of a node with other sets of nodes, are called **co-affiliation** [38]. Examples of co-affiliations are 'attendance at the same event', 'membership in the same club, and members of the same corporate board'.

To obtain the co-affiliation in the network, the affiliation network is transformed from a two-mode network into a one-mode network (network with only one type of node). This procedure of transformation is called **projection** [37]. Projection in affiliation networks is carried out by selecting one of the sets of nodes and linking two nodes from that set if they were connected to the same node of the other set. This means that projection allows the analysis of the network from one of two perspectives: the actor's view or the event's view [267]. The procedure of projection is also referred to as the duality of the two-mode network since it allows dual perspectives (one from each mode) of the affiliation network. From the actor's viewpoint, two actors are connected if they have participated in at least one event together, and from the event's viewpoint, two events are connected if at least one of the actors has participated in these events.

4.5.1   Representing Affiliation Network

One of the best ways to represent networks is through a matrix. A matrix is a rectangular table in which rows and columns represent the nodes of the network and

Table 4.1 Affiliation matrix of author-paper affiliation network

	Paper 1	Paper 2	Paper 3	Paper 4
Author 1	1	0	1	0
Author 2	1	1	0	1
Author 3	0	1	0	1

the value in the cell (where the column and row intersect) represents an edge. Similarly, affiliation networks are represented through an affiliation matrix, $A = \{a_{ij}\}$. Matrix A is a two-mode sociomatrix in which rows represent actors and columns represent events. Generally, affiliation network A is defined as: A is a bipartite graph $A = (U, V, E)$ where U (often known as actors) and V (often known as events) are a disjoint set of nodes and $E \cup (U X V)$ is the set of edges. With $p = |V|$ and $q = |U|$, A is represented by an incident matrix with p lines and q columns. Formally, $A = \{a_{ij}\}$ records the affiliation of each actor with each event in an affiliation matrix such that:

$$a_{ij} = \begin{cases} 1, & if \ actor \ i \ is \ affiliated \ with \ event \ j \\ 0, & otherwise \end{cases} \tag{4.3}$$

The value of 1 is put in the (i, j)th cell if ith actor (ith row) is affiliated with jth event (jth column) and an entry of 0 if ith actor is not affiliated with jth event. Table 4.1 represents the affiliation matrix of a sample author-paper affiliation network. This is a bipartite graph with two types of nodes, namely authors and papers. The edge (link) in the network shows which authors have written which papers and two authors linked to the same paper represent a co-authorship relationship, as depicted in Fig. 4.9. For example, in Fig. 4.9, author 1 has written two papers, numbered 1 and 3. Paper 3 has only one author, namely author 1, however for paper 1, author 2 is a co-author with author 1.

4.5.2 Projecting Affiliation Network

As mentioned above, it is possible to analyse a two-mode network in its original form; however. few methods exist for this purpose. Often, two-mode networks are transformed into a one-mode network by a procedure called projection. Projection generates a one-mode network by selecting the nodes of one set (for example, authors in Fig. 4.9) and linking two nodes from the set if both are connected to the same node of the other set. The projection of a two-mode network into two one-mode networks provides the opportunity to analyse the affiliation network using methods developed for traditional unipartite or social networks. The transformed one-mode (co-affiliation) network helps the understanding and analysis of the ties between the

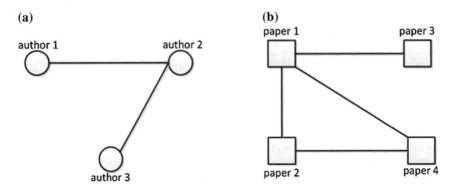

Fig. 4.10 Example of projection: **a** authors co-affiliation network, **b** paper's co-authorship network

members of a node set [38]. For example, Fig. 4.10a presents the authors co-affiliation network and Fig. 4.10b presents the papers co-authorship network.

In the next section, the Ontology Usage Network Analysis Framework (OUN-AF) which is proposed for the study and analysis of the relationships between different ontologies and data sources in a dataset using SNA techniques is presented.

4.6   Ontology Usage Network Analysis Framework (OUN-AF)

The objective of the ontology identification phase is to identify the use of different ontologies by different data publishers in a given application area to discover hidden usage patterns. To mine such analysis, the Ontology Usage Network Analysis Framework (OUN-AF) is proposed, as shown in Fig. 4.11. OUN-AF comprises three phases: *Input* phase, *Computation* phase and *Analysis* phase, as shown in Fig. 4.11. The role of the *Input* phase is to collect and maintain the dataset containing the crawled data, which is comprised of real world Semantic Web data. The *Computation* phase provides the computational architecture to transform the input into a format that facilitates ontology identification-related activities. The *Analysis* phase analyses the computational model by using the developed metrics and interprets their results. Each phase is discussed in detail in the following subsections.

4.6.1   Input Phase

The input phase is responsible for managing the dataset which is then used for subsequent operations. The two key components in this phase are a *crawler* and an *RDF triple store*. The crawler crawls the Web to collect the required data to form

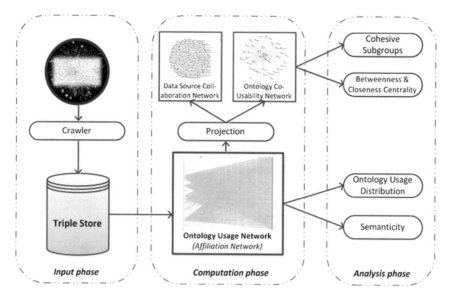

Fig. 4.11 Ontology usage network analysis framework (OUN-AF) and its phases

a dataset which is then stored in the RDF triple store. The bootstrapping process first builds the seed URIs as multiple starting points to point the crawler to relevant and interesting data sources. A list of seed URIs is obtained by accessing semantic search engines which return the URIs (URLs) of the data sources (web sites) with structured data. The crawler collects the Semantic Web data (RDF data) and, after preprocessing it, loads it to the RDF triple store (database).

Preprocessing handles the transformation routines for converting the crawled data into the required format and appending the necessary metadata such as provenance detail, timestamp, and data source details. The Named Graph [53] approach is used for data management, since RDF data comprises triples (statements) which do not provide a default mechanism to group or associate certain sets of triples with a context. The Named Graph approach enables the provision of contextualization by introducing an additional URI (context) to a set of related triples.

The dataset is then accessed by the components of the computation phase to query the information. SPARQL end point is used to pose SPARQL queries, which access the triple store to evaluate the query and return the result set. The result set, in this case, is the set of RDF triples in the form of an RDF graph.

4.6.2 Computation Phase

The computation phase provides the computational architecture to transform the data maintained in the RDF triple store to a model so that further analysis can be performed on the given dataset and ontologies and their usage patterns can be identified. The computational architecture comprises a model to represent the ontologies and their relationship with data publishers and a network operation in a network-based structure to analyse the ontologies, their interrelationship with other ontologies and relationship with data publishers. The OUN-AF transforms the data into two formats. The first format is a two-mode affiliation network and the second format is a one-mode network which is generated from a two-mode network using the projection procedure. The two-mode affiliation network (i.e. *Ontology Usage Network*) and the consequent one-mode networks (i.e. *Ontology Co-Usability* and *Data-Source Collaboration* networks) are discussed in the following subsection.

4.6.2.1 Ontology Usage Network (OUN)

OUN is an affiliation network represented as a bipartite graph which provides the model that allows the creation of a relationship between two distinct sets of nodes and the analysis of the use of ontologies by different data publishers. OUN comprises *ontology* and *data-source* sets of nodes with an edge between the ontology node and data-source node if the ontology has been used by the data source. Here, data source refers to any domain name on the Web which has published RDF data using Web ontologies. To formally define the Ontology Usage Network, the two sets of nodes *ontology set* and *data-source set* are first defined and the OUN definition is then presented.

An ontology set is defined as the set O which represents the nodes of the first mode of the affiliation network. An ontology set O contains the list of ontologies used on the Web-of-Data such that there is a triple $t < s, p, o >$ anywhere in the dataset (specifically, otherwise in general, on the Web-of-Data) where $o \in O$ is the URI of either p or o.

A data-source set is defined as the set D which has the list of hostnames on the Web-of-Data such that there exists a triple $t < s, p, o >$ in the dataset (specifically, otherwise in general, on the Web-of-Data), where s is the hostname (domain names in URL parlance) and either p or o is a member of O.

The Ontology Usage Network (OUN) is a bipartite graph, denoted as OUN (O, D), which represents the affiliation network with a set of ontologies O on one side and a set of data sources D on the other, and edge (o, d) represents the fact that o is "used" by d. A snapshot of OUN is shown in Fig. 4.12

There are certain types of analysis which cannot be obtained directly through OUN, particularly if the requirement is to infer the connectedness in one set of nodes based on their co-participation in the other set of nodes. For this kind of analysis, it is necessary to study one set of nodes, hence the information represented by OUN has to be transformed from a two-mode network to a one-mode network. This transformation is achieved using the process of projection discussed next.

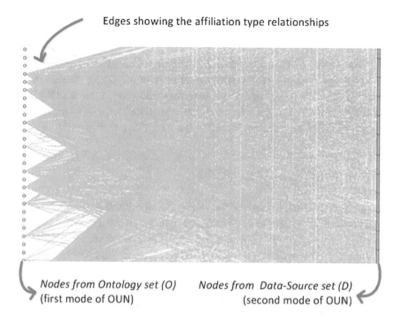

Fig. 4.12 Ontology usage network (affiliation network with one set of nodes representing ontologies and the other set of nodes representing data sources)

4.6.2.2 Projection

OUN is a two-mode network which enables the study of the bipartism found in the network, however it is sometimes desirable to obtain one set of nodes and study their co-membership in the network. This is achieved by transforming a two-mode network into a one-mode network using a technique called projection. In the case of OUN, projection is used to generate two one-mode graphs; one for nodes in the ontology set known as the *Ontology Co-Usability* network (See Fig. 4.19), and one for the data source set known as the *Data-Source Collaboration* network (See Fig. 4.18).

The Ontology Co-Usability network is an ontology-to-ontology network, in which two nodes are connected if both ontologies are being used by the same data source. This means that the *Ontology Co-Usability* network represents the connectedness of an ontology with other ontologies, based on their co-membership in the data source.

The Data-Source Collaboration network is a data-source-to-data-source network in which two nodes are connected if both of them have used the same ontology to describe their data. The Data-Source Collaboration network represents the similarity of data-sources in terms of their need to semantically describe the information on the Web.

4.6.3 Analysis Phase

The analysis phase is the third and last phase of OUN-AF. The objective of this phase is to mine the hidden relationships explicitly or implicitly present in the two-mode network (i.e. OUN) and one-mode networks (Ontology Co-Usability and Data-Source Collaboration). To objectively analyse the networks, the SNA techniques used and the metrics developed are explained in the next two subsections.

4.6.3.1 Analysing (Two-Mode) OUN

Two metrics are proposed to analyse the OUN. The quantitative analysis on the OUN affiliation network provides the infrastructure to measure the degree of nodes in each set of modes. In the case of OUN, the degree of the nodes representing ontologies and the degree of the nodes representing data sources is obtainable. Additionally, the degree distribution, which is the probability distribution of node degrees over the whole network, can be obtained to compare the network and its connections with other types of networks. To obtain the degree and degree distribution of the OUN, two metrics are defined:

- *Ontology Usage Distribution*: this measures the degree of ontologies and their distribution over the network.
- *Semanticity*: this measures the degree of the data sources and their distribution over the network.

These two metrics are formally described in Sects. 4.7.1 and 4.7.2, respectively.

4.6.3.2 Analysing Projected One-Mode Network

As mentioned above, the projection procedure is applied to transform the OUN into two projected one-mode networks. Each resultant projected network contains nodes from their own set (e.g. ontologies) and the relationship between nodes shows their co-affiliation in the original two-mode network. These networks provide the foundation for discovering other interesting properties which further aids the understanding of how ontologies are placed in terms of their usage and co-usage by different data sources and their typological position within the network. The two networks obtained are the Ontology Co-Usability network, which is essentially an ontology-to-ontology network, and the Data-Source Collaboration network, which is a data-source-to-data-source network.

As the focus of the analysis phase of OUN-AF (and of OUSAF as well) is to analyse ontology usage, only the ontology-to-ontology network is considered. The ontology-to-ontology network represents the relationships between the different ontologies in a dataset. SNA provides the techniques and methods to study the strategic position of nodes in the overall network. To obtain this insight, understand the position of the nodes and the groups of nodes with similar positions (which form the cluster), the following metrics are used:

- *Betweenness and Closeness Centrality*: identify the nodes which have important strategic positions in the network such as betweenness and closeness
- *Cohesive Subgroups*: identifies the group of nodes which share some similarities, particularly in terms of their relationship and position

These two metrics are formally described in Sects. 4.7.3 and 4.7.4, respectively. It is important to note that the above-mentioned metrics can also be used to analyse the Data-Source Collaboration network as well, if required.

In the next subsection, the set of sequential activities carried out in the OUN-AF to analyse the network are summarized.

4.6.4 Sequence of OUN-AF Activities

OUN-AF comprises three phases, and a different set of activities is involved in each phase. A summary of key activities mentioned below provides an overview of the activities and their sequence.

- In the Input phase, the dataset is built. To build the dataset comprising the information on the use of ontologies by different data publishers, sources for crawling the data are identified.
- In the computation phase, the crawled data is processed to extract the relevant information to build the two node sets of OUN, namely *ontology set* and *data-source*.
- To study the two-mode network, using these two set of nodes, the OUN is constructed. The affiliation relationship between these two sets of nodes is established, based on the usage-related data represented in the dataset.
- Metrics are developed to perform the required analysis. They are:

  - Ontology Usage Distribution (OUD)
    First, the degree of each node that is a member of the *ontology set* is measured. Second, the degree distribution of the *ontology set* is measured to understand the distribution of degrees in the node set, and to understand the distribution of connections in the network. Power-law distribution is observed in the network.

  - Semanticity
    First, the degree of each node that is a member of the *data-source set* is measured.
    Second, the degree distribution of the *data-source set* is measured to understand the distribution of degree in the node set.

- To study the one-mode network, the OUN network is transformed into two one-mode networks using the projection operation to produce the *Ontology Co-Usability* and *Data-Source Collaboration* networks.

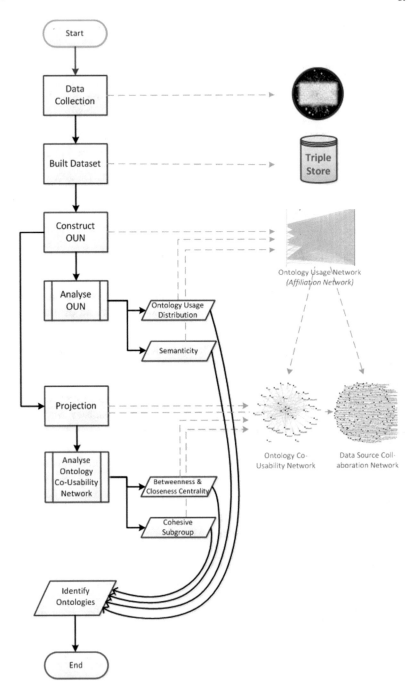

Fig. 4.13 Flow of activities in OUN-AF

- To measure the position and identify the key nodes in the network, the following two metrics are used on projected one-mode networks (i.e. Ontology Co-Usability)

  - Betweenness and closeness centrality measures that identify the nodes in key strategic positions.
  - Cohesive subgroups that help to identify the clusters in the network, based on functional or structural similarities.

- Interpret the results and identify the ontologies based on the required selection criteria

The above-mentioned set of activities is depicted in Fig. 4.13, based on workflow notations highlighting the key activities and their corresponding outputs. The output generated as the results of the activity or process is shown using a grey dotted line, whereas the set of activities follows normal workflow representation.

In the next section, the metrics used to study the OUN and the projected one-mode network as part of OUN-AF are defined.

4.7 Metrics for Ontology Identification

The metrics used for the identification phase are presented in the following subsection.

4.7.1 Ontology Usage Distribution (OUD)

The metric Ontology Usage Distribution, OUD_k is used to identify which fraction of the ontologies in the network have a degree k. Recall that in an ordinary graph, the degree of a node in a network is the number of edges connected to that node. However, in the case of an affiliation network, the degree of a node is the number of ties it has with the number of nodes of the other set. The degree centrality $C_D(o_i)$ of an ontology o_i is measured as:

$$C_D(o_i) = d(o_i) = \sum_{j=1}^{n_1} A_{ij} \tag{4.4}$$

where $i = 1, \ldots n_1$, $n_1 = |O|$, $d(o_i)$ is the degree of i_{th} ontology o, and A is the affiliation matrix representing OUN.

The normalization (which aligns the probability distribution to an adjusted value) of degree in an affiliation network is obtained by dividing the total number of nodes in the other set rather than dividing by the number of nodes in the same set. Therefore, the normalized ontology usage degree is measured as:

$$C'_D(o_i) = \frac{d(o_i)}{number\_of\_datasources - 1} \tag{4.5}$$

where $number\_of\_datasources = |D|$ represents the total number of nodes in the other set of nodes (i.e. 2nd mode of affiliation network).

4.7.2   Semanticity

Semanticity distribution $Semanticity_k$ identifies the fraction of the data sources in the network which have a degree k. Similarly to ontology usage distribution which measures the distribution of ontologies among data sources, semanticity measures the participation of different ontologies in a given data source. In other words, semanticity measures the richness of a given data source in terms of the use of ontologies. The more ontologies are used by a data source, the higher semanticity value the source has. Semanticity is measured by calculating the degree centrality and degree distribution on the second set of nodes in an affiliation network, which is the set representing the data sources in the dataset. The degree centrality $C_D(ds_i)$ of a data source ds_i is measured as:

$$C_D(ds_i) = d(ds_i) = \sum_{j=1}^{n_2} A_{ij} \tag{4.6}$$

where $i = 1, \ldots n_2$, $n_2 = |D|$, $d(ds_i)$ is the degree of i_{th} data source ds, and A is the OUN

The normalization of Semanticity is measured as:

$$C'_D(ds_i) = \frac{d(ds_i)}{number\_of\_ontologies - 1} \tag{4.7}$$

where $number\_of\_ontologies = |O|$ represents the total number of nodes in the other set of nodes (i.e. the second mode of an affiliation network).

4.7.3   Betweenness and Closeness Centrality

Betweenness and Closeness centrality is used to identify the important (or key) nodes in the network. Like Ontology Co-Usability, both these centrality measures are computed on the projected one-mode network.

Betweenness Centrality is the number of shortest paths between any two nodes that pass through the given node. The betweenness centrality $C_B(v_i)$ of a node v_i is measured as:

$$C_B(v_i) = \sum_{v_s \neq v_i \neq v_t \in V} \frac{\sigma_{st}(v_i)}{\sigma_{st}} \qquad (4.8)$$

where σ_{st} is the number of shortest paths between vertices v_s and v_t.

$\sigma_{st}(v_i)$ is the number of shortest paths between v_s and v_t that pass through v_i.

Closeness centrality is a measure of the overall position of a node (actor) in the network which provides an indication of how long it takes to reach other nodes in the network from a given starting node. Closeness measures reachability, that is, how fast a given node (actor) can reach everyone in the network [210] and is defined as:

$$c_v = \frac{n-1}{\sum_{u \in V} d(v,u)} \qquad (4.9)$$

where $d(v,u)$ denotes the length of the shortest-path between v and u. The closeness centrality of v is the inverse of the average (shortest path) distance from v to any other vertex in the graph. The higher the c_v, the shorter the average distance from v to other vertices; v is more important by this measure.

4.7.4 Cohesive Subgroups

Generally speaking, cohesive subgroups refer to the areas of the network in which actors are more closely related to each other than to actors outside the group. In the most extreme case of a cohesive subgroup, each member of the group is expected to have strong connections with every other member of the group. However, this condition is very strict and is normally relaxed by introducing the notion of cliques or n-clique. In an n-clique, it is not required that each member of the clique has a direct tie with others, rather that it has to be no more than distance n from another.

Formally, a clique is the maximum number of actors that have all possible ties between them.

In the next section, the analysis and results obtained using these metrics to identify the ontologies in a real world data set are discussed.

4.8 Analysis of Ontology Usage Network

To base the findings on empirical grounds, a dataset comprising real world instance data is built to populate the OUN and, using the metrics described in the earlier section, analysis is performed to understand the relationships between the ontologies, data publishers and the interrelationships between the ontologies based on their co-usage by different data sources.

edges

(first mode)
Ontologies (44)

(second mode)
Data sources (211)

Fig. 4.14 Ontology usage affiliation network (bipartite graph)

4.8.1 Dataset and its Characteristics

A dataset comprising real world structured data which is annotated using ontologies is developed for the *identification* phase. In order to build a dataset which has a fair representation of the Semantic Web data described using domain ontologies, semantic search engines such as Sindice [266] and Swoogle [79] are used to build the seed URLs. These seed URLs are then used to crawl the structured data published on the Web using ontologies. The dataset built for the identification phase comprises 22.3 million triples, collected from 211 data sources.[2] In this dataset, 44 namespaces are used to describe entities semantically. The resulting Ontology Usage Network is depicted in Fig. 4.14 and comprises 1390 edges linking 44 ontologies to 211 data sources. The complete list of ontologies and their prefixes used in the dataset collected by crawling the Web is shown in Fig. 4.15.

---

[2]https://docs.google.com/spreadsheet/ccc?key=0AqjAK1TTtaSZdGpIMkVQUTRNenlrTGctR2
J1bkl6WEE.

#	Prefix	Namespace	Degree
1	rdf	http://www.w3.org/1999/02/22-rdf-syntax-ns#	208
2	gr	http://purl.org/goodrelations/v1#	208
3	rdfs	http://www.w3.org/2000/01/rdf-schema#	190
4	vCard	http://www.w3.org/2001/vcard-rdf/3.0# & http://www.w3.org/2006/vcard/ns#	164
5	owl	http://www.w3.org/2002/07/owl#	141
6	foaf	http://xmlns.com/foaf/0.1/#	115
7	xhtml	http://www.w3.org/1999/xhtml/vocab#	126
8	dc	http://purl.org/dc/	75
9	eClass	http://www.ebusiness-unibw.org/ontologies/eclass/5.1.4/#	38
10	v	http://rdf.data-vocabulary.org/#	35
11	ogp	http://ogp.me/ns# & http://opengraphprotocol.org/schema/	18
12	sindice	http://vocab.sindice.net/date	16
13	rev	http://purl.org/stuff/rev#	15
14	yahoo	http://search.yahoo.com/searchmonkey/commerce/Business	11
15	pro	http://www.productontology.org/id/	4
16	wgs84	http://www.w3.org/2003/01/geo/wgs84_pos#	2
17	fb	http://www.facebook.com/2008/fbmladmins	2
18	powder	http://www.w3.org/2007/05/powder-s#	1
19	inhouse1	http://herbaman.com.ar/Products.html	1
20	inhouse2	http://lokool.com/extendedgoodrelations.owl	1
21	inhouse3	http://www.kica-jugendstil.com/semanticweb.rdf	1
22	inhouse4	http://www.logicpass.com/semanticweb.owl	1
23	inhouse5	http://www.openlinksw.com/schemas/DAV	1
24	inhouse6	http://www.acigroup.co.uk/semanticweb.rdf	1
25	inhouse7	http://www.buntegeschenke.de/semanticweb.rdf	1
26	inhouse8	http://www.svanvit.se/sv/kvinna/shopkvinna	1
27	inhouse9	http://www.symbolontarot.nl/de-winkel-met-symbolon-artikelen.html	1
28	inhouse10	http://data.openlinksw.com/oplweb	1
29	inhouse11	http://olutools.com/shop.html	1
30	inhouse12	http://www.wifo-ravensburg.de/rdf/semanticweb.rdf	1
31	comm	http://purl.org/commerce#	1
32	coo	http://purl.org/coo/ns#	1
33	media	http://purl.org/media#	1
34	scovo	http://purl.org/NET/scovo#	1
35	vso	http://purl.org/vso/ns#	1
36	void	http://rdfs.org/ns/void#	1
37	sioc	http://rdfs.org/sioc/ns#	1
38	frbr	http://vocab.org/frbr/core#	1
39	cc	http://creativecommons.org/ns# & http://web.resource.org/cc/license	0
40	vann	http://purl.org/vocab/vann/	0
41	skos	http://www.w3.org/2004/02/skos/core#	0
42	vocab	http://www.w3.org/2003/06/sw-vocab-status/ns#	0
43	rdfa	http://www.w3.org/ns/rdfa#	0
44	g	http://www.w3.org/2003/g/data-view#	0

Fig. 4.15 List of ontologies with their prefixes

In terms of generic OUN properties, the *density* of the network is 0.149 (Eq. 4.2) and the *average degree* is 10.90 (Eq. 4.1). The average degree shows that the network is neither too sparse nor too dense, which is a common pattern in information networks. Details on the other properties and metrics are given in the following subsections.

4.8.2   Analysing Ontology Usage Distribution (OUD)

Ontology Usage Distribution (OUD) refers to the use of ontologies by data sources in publishing their information and can be used to determine *how the use of an ontology is distributed over the data sources in the dataset*. The Ontology Usage Network is analysed to measure the degrees of the nodes.

Observation: Using Eqs. 4.4 and 4.5, Table 4.2 shows the percentage of the ontologies being used by a number of data sources. The relative frequency of OUD on the dataset shows that there is both extreme and average ontology usage by data sources. It also shows that 13.6% of the ontologies are not used by any of the data sources and approximately half of the ontologies are used exclusively by the data sources. The second row of the Table 4.2 shows that 47.7% of the ontologies (21 ontologies) are being used by a data source that has not used any other ontology. This means that there are several ontologies in the dataset which either conceptualize a very specialized domain, restricting their reusability, or are of a proprietary nature. From the third row of Table 4.2 onwards, there is an increase in the reusability factor of ontologies. This is because they are being used by an increasingly large number of data sources. The last row shows that 4.5% (two ontologies) of ontologies are being used by 208 data sources. Through this analysis, it can be seen that there are fewer ontologies which are not being used at all and a small number which have almost optimal utilization.

Figure 4.15 shows the complete list of ontologies used in the dataset along with their degree (the number of data sources using the ontology). As previously mentioned, vast numbers of ontologies are used by only one data source, which indicates that they are either very specialized in nature and/or are proprietary for exclusive use. In Fig. 4.15, rows 18 to 38 show the namespaces of these ontologies, some of which cover very specific domains and some of which are proprietary. Although the licence terms of ontologies assumed to be proprietary were not found, the non-availability of their specification document leads us to believe this is the case.

Figure 4.16 shows the degree distribution of ontology usage in a number of data sources. The value of degree is shown on the x-axis and the number of ontologies with that degree is shown on the y-axis. It can be seen that a large number of ontologies have a small degree value and only a few ontologies have a larger degree value.

Table 4.2 Distribution of ontology usage in data sources

# of data sources	# ontologies	% ontologies
0	6	13.6
1	21	47.7
2	1	2.3
3	1	2.3
4	1	2.3
11	1	2.3
15	1	2.3
16	1	2.3
18	1	2.3
38	1	2.3
75	1	2.3
115	1	2.3
126	1	2.3
141	1	2.3
164	1	2.3
190	1	2.3
208	2	4.5

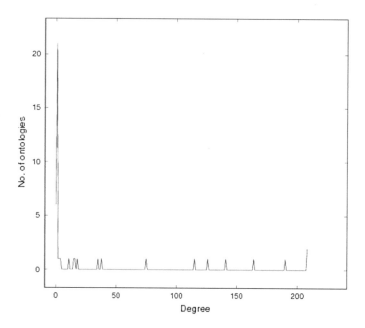

Fig. 4.16 Degree distribution of ontology usage (data sources per ontology)

Table 4.3 Distribution of semanticity (Ontology used per data source)

# ontologies	# of data source	% data source
2	1	0.5
3	4	1.9
4	3	1.4
5	38	18.0
6	59	28.0
7	51	24.2
8	39	18.5
9	13	6.2
10	2	0.9
14	1	0.4

4.8.3 Analysing Semanticity

Semanticity measures the richness of a data source in terms of ontology usage. By semanticity, we mean the ability of any data source (data publisher) to provide semantically rich structured data that is annotated by one or more ontologies. The assumption is that the higher the number of ontologies used by the dataset, the more semantically rich the data source must be. Semanticity, which is essentially the number of ontologies per data source, is obtained by measuring the degree of the nodes of the data source in the Ontology Usage Affiliation network.

Observation: Using Eqs. 4.6 and 4.7, in the OUN, it is observed that on average, 6.6 ontologies per data source are used in the dataset which, in our view, is an encouragingly high semanticity value, particularly bearing in mind that there are several ontologies with very low ontology usage degree values such as 0 and 1, as described in the previous section on Ontology Usage Distribution. After determining the average semanticity (i.e. Eq. 4.7) of the data sources, their degree distribution is observed. Table 4.3 shows the relative frequency of ontologies being used by a number of data sources. The degree distribution of ontology usage per data source is different from ontology usage distribution.

At the lowest level, two ontologies in the network[3] are used by one data source, which shows the lowest semanticity value, while 14 is the maximum semanticity value which is also used by one data source. When the degree distribution of the data source is plotted, it follows the Gaussian distribution shown in Fig. 4.17. Gaussian distribution [285], which is essentially a bell-shaped curve, is normally concentrated in the centre and decreases on either side. This signifies that degree has a lesser tendency to produce extreme values than power law distribution. It is believed that Gaussian distribution (also known as normal distribution), which circumvents the

---

[3] www.oettl.it and www.openlinksw.com.

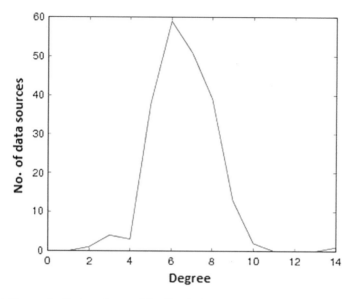

Fig. 4.17 Degree distribution of semanticity (Ontologies per data source)

exponential growth in degree distribution, is quite helpful in designing algorithms that need to consume data on the Web from a scalability point of view.

Let us now look at each set of nodes separately to analyse their characteristics and better understand the emerging relationships in the nodes of the same set. Two networks from the Ontology Usage Affiliation network are generated to do this, using the projection process discussed in Sect. 4.5.2, namely the Ontology Co-usability network (Fig. 4.19) and Data-Source Collaboration network (Fig. 4.18).

In the following section, we focus only on the Ontology Co-usability network and analyse its properties in detail.

4.8.4 Formation of Ontology Co-usability

Ontology co-usability is an undirected graph extracted by projecting the Ontology Usage Affiliation network onto an ontology set of nodes to form an ontology-to-ontology network. In an ontology co-usability graph, ontologies are linked to other ontologies if they are being used by the same data source. Collaboration networks such as the projected Ontology Co-usability network are of coarser representation than the affiliation network; however, they are still more informative, since many collaboration patterns are available through these graphs [101], such as components and cohesive subgroups [99].

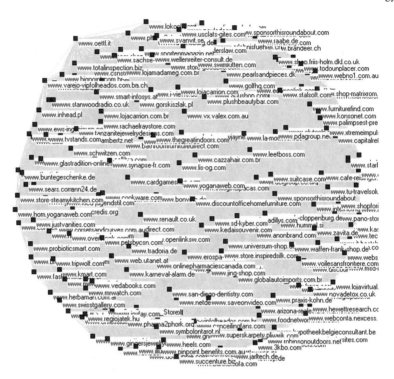

Fig. 4.18 Data source collaboration network

Observation: In the dataset being analysed, the Ontology Co-usability network comprises 44 vertices, and there are 305 edges between these vertices (Fig. 4.19). The network has 38 loops which indicate the affiliation of nodes in the network, thus, this tells us that six nodes (which means six ontologies, that is, nodes without any edges in Fig. 4.19) are not being used by any data source at all. For general network properties, the density and average degree (Eqs. 4.2 and 4.1 respectively) of the Ontology Co-usability network is 0.295 and 13.86. It is interesting to see that although we tend to lose some information through the projection process (two-mode to one-mode), the extracted network is denser (Ontology Usage Affiliation density is 0.149) and has a high occurrence of cliques. Also, the average degree of ontology co-usability is 13.86, which is higher than the original bipartite graph, i.e. 10.90. This shows that a large number of ontologies are mutually (collaboratively) used by different data sources having data which is common or related in nature and semantically similar to each other. This highlights the fact that we can generate a minimum set of vocabularies (URIs) of the interlinked ontologies, representing the schema requirements of publishers, to facilitate querying and inferencing information efficiently on the Web.

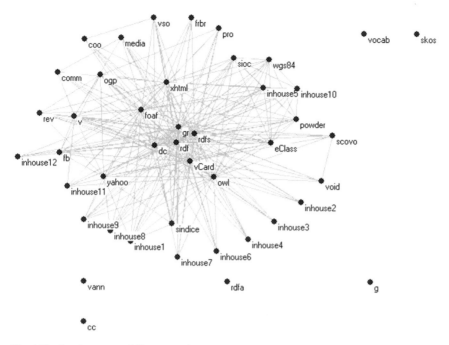

Fig. 4.19 Ontology co-usability network

In the next section, we will look at the centrality of the vertices of the Ontology Co-usability network to understand which ontologies are more central in terms of betweenness and closeness measures.

4.8.5 Analysing Betweenness and Closeness

As mentioned earlier, betweenness centrality measures the number of shortest paths to go to a certain node. This is based on the notion that a node which lies between the shortest paths of many nodes has a central position in the network. Nodes with high betweenness centrality are important for other nodes to reach (communication gateway) since they fall on the geodesic paths between other pairs of nodes.

On the other hand, the closeness centrality of a node in the network is the inverse of the average shortest path distance from the node to any other node in the network [200]. The larger the closeness centrality of a node, the shorter the average distance from the node to any other node; thus, this is interpreted as the nodes efficiency in spreading information to all other nodes.

Observation: In the context of the Ontology Co-usage network, the interpretation of betweenness and closeness measures are different from other collaboration networks such as co-authorship collaboration networks [201]. In betweenness, which is measured using Eq. 4.8, the nodes of larger values are considered to be the hub of the network, controlling the communication flow (or becoming the major facilitator) between nodes with a geodesic path passing through this hub. In the Ontology Co-usage network, ontologies are linked based on the fact that they are being co-used by the data sources, therefore it is believed that the ontologies with maximum between-ness centrality act as a semantic gateway[4] and becomes a major motivational factor for the usage of other ontologies.

Likewise, in closeness centrality which is measured using Eq. 4.9, the larger the value, the shorter the average distance from the node to any other node, and thus the node (with a larger value) is positioned in the best location to spread information quickly [209]. This centrality measure in the ontology co-usage graph enables the establishment of correspondence between ontologies which have concepts related to each other, supplementing each others conceptual model to form an exploded domain. The utilization of ontology indexing based on closeness centrality is very similar to the features discussed in [71] in supporting the application specific use of ontologies such as:

(i)   the ontologies closer to each other in their usage are better candidates for vocab-ulary alignment,

(ii) ontologies closer to each other have more entities which correspond to entities of other ontologies, and

(iii) closely related ontologies tend to facilitate query answering on the Semantic Web.

The betweenness and closeness centrality of ontology co-usage nodes is shown in Figs. 4.20 and 4.21, respectively. The node size in Figs. 4.20 and 4.21 reflects the centrality value.

As can be seen in Fig. 4.20, the Ontology Co-Usage network has very few nodes with a higher betweenness value, which means that the ontologies represented by the green nodes (which are few) are the ones falling in between the geodesic path of many other nodes and acting as the gateway (or hub) in the communication between other ontologies in the graph. These are the nodes, namely `rdf`, `rdfs`, `gr`, `vCard` and `foaf` which, in our interpretation, act as the semantic gateway by becoming the reason for the adoption of other ontologies on the Web. On the other hand, closeness centrality is approximately distributed evenly in the network. Thus, it is safe to assume that almost every node is reachable by other nodes except for those which are not connected.

---

[4]Semantic Gateway can be considered as the Drug Gateway effect, in which a certain drug becomes the driving force (or reason) for the utilization of other drugs.

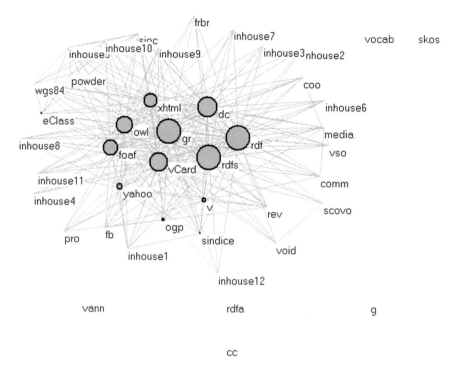

Fig. 4.20 Betweenness centrality of ontology co-usage network

4.8.6 Analysing Cohesive Subgroups

A connected component in an undirected graph is a sub-graph in which any two nodes are connected to each other by a path, traversable through intermediate nodes. In a collaboration graph such as the Ontology Co-Usage network, a connected component is a maximal set of ontologies that are mutually reachable (and connected) through a chain of co-usage links. The connected components reveal the state of connectedness of the ontologies in the Semantic Web in general and in our dataset specifically [130]. It is believed that to promote the reusability of knowledge and allow several conceptual models to interplay on the Web, a cohesive subgroup of ontologies formed by widely connected components is a desirable property.

Observation: A cohesive sub-group analysis to identify connected components of the Ontology Co-Usage network shows that the network is widely connected. The connected component is 86.36% (See Fig. 4.22; only six are not connected in the network (this means 0-core), while others are connected with varying k-core values), making it a giant network since it encompasses the majority of the nodes. This means that 86.36% of the ontologies are reachable via one another by following the links (domain names URIs) of the data sources included in the dataset. Note

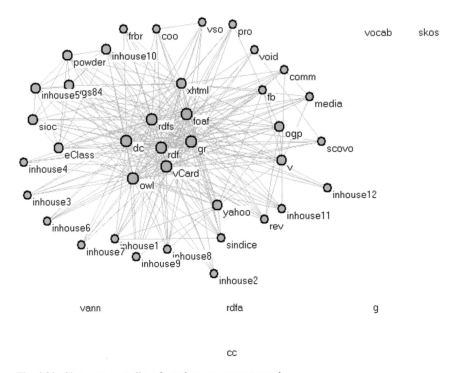

Fig. 4.21 Closeness centrality of ontology co-usage network

that the size of the cohesive sub-group, in terms of percentage, closely matches the findings of [48] for the classical Web which was 91%.

To know the sub-component within the giant connected component, based on the equal distribution of the concentration of links around a set of nodes, k-core is computed. k-core is the maximum sub-graph in which each node has at least degree k within the sub-graph. Figure 4.22 stacks the k-core components, based on ascending k values from highest to lowest. From Fig. 4.22, it is easy to see which ontologies are highly linked, based on the invariance in ontology usage patterns across data sources.

4.9   Ontology Identification Evaluation

The aim of the ontology identification phase is to identify the ontologies which can be analysed to understand their usage patterns and trends in detail. The OUN-AF provides a model and analysis that is capable of addressing the selection criteria requirements, mentioned in Sect. 4.3. How the OUN-AF can assist in identifying different ontologies of interest, according to the selection criteria in two different scenarios, is demonstrated in the next subsection.

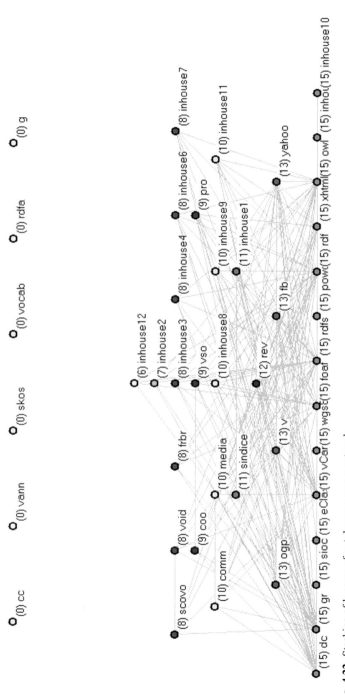

Fig. 4.22 Stacking of k-cores of ontology co-usage network

4.9.1   Scenario 1: Ontologies and Data Publishers

Let us consider two scenarios:

Case 1: From the given dataset, analyse how many data publishers (ontology users) are using a given ontology to describe their domain-specific entities.

Analysis: A more generalized description of this requirement is to understand the distribution of ontologies over the data -sources included in the dataset. To obtain answers for such queries, the *Ontology Usage Distribution* (OUD) metric (Sect. 4.7.1) is used to measure the degree and its distribution over the OUN. Figure 4.15 lists all the ontologies and their degree values. The degree value of each ontology tells the number of unique data publishers (data sources) using them. By examining the list, it can be seen that in the dataset (which focuses on the e-Commerce application area) *gr, vCard* and *foaf* are being used by 208, 190 and 115 data sources, respectively.

Additionally, the degree distribution plot, shown in Fig. 4.16, helps us to understand how ontologies are generally being adopted. It is also observed that the degree distribution, as shown in Fig. 4.16, closely follows the power law distribution, albeit not exactly, which, in fact, is the distribution model observed in several information networks, particularly internet (or Web) networks [2]. Networks with power-law distribution are also sometimes referred to as scale-free networks [15] because they tend to be scale-free. Looking at the distribution and following the patterns found in such scale-free networks, it can be safely assumed that the use of ontologies by different data sources follows "preferential attachment" [14].

Case 2: What other ontologies are being used by the same data publisher to understand the level of semanticity present?

Analysis: To understand what other vocabularies are being co-used by a data publisher to semantically describe the entities representing the application area, *Semanticity* metrics (Sect. 4.7.2) are used. From the results obtained based on the dataset used, it can be seen that the top 10 vocabularies being co-used by data sources with the highest semanticity value in the eCommerce domain (including standard vocabularies) are: *rdf, gr, rdfs, vCard, owl, foaf, xhtml, dc, eCl@ss* and *v*.

It is observed that the distribution follows Gaussian distribution, as shown in Fig. 4.17, which means that a few of the data sources have higher semanticity than the others, making it easy to identify which data sources are publishing semantically rich structured data.

4.9.2   Scenario 2: Ontologies Co-usability

It is always desirable and somewhat interesting to know how things are interlinked and co-used on the Web. Ontology co-usability, which is produced through the projection procedure over the affiliation network, helps to identify which ontologies are being co-used with other ontologies and with what frequency. This means that the link between two nodes (ontologies) in the one-mode network, produced through projection, tells us that these two ontologies are co-jointly being used to describe information by a data source. From Fig. 4.19, it is clear that except for six ontologies, all are being co-used, which is a positive trend and helps in realizing the Semantic Web vision where entities are not only semantically described but also interlinked with other entities to form the *Web-of-Data*, which in turn is processable by computers. Based on the analysis, ontologies which are largely co-used with other ontologies are *gr, foaf, dc,* and *vCard.* Here, W3C-based vocabularies are not discussed because firstly, these are meta-languages and secondly, they do not represent a domain or application area.

4.10   Recapitulation

In this chapter, the OUN-AF that assists in the identification phase of the OUSAF was presented. The OUN-AF and its components (dataset, OUN and metrics) enable a detailed insight into how different data sources are using particular ontologies and how they are being co-used. The analysis will also assist in understanding how different ontologies are interlinked and their usage patterns in the dataset. This insight, based on real instance data obtained by crawling the Web, provides substantial evidence as to how different but related domain-specific ontologies are being co-used by data publishers to provide semantically rich structured data on the Web. The identification of prominent domain ontologies helps data publishers and application developers to achieve a better experience.

In the next chapter, the *investigation phase* of the OUSAF is presented. The output of this phase i.e. identification of ontologies, provides the input to perform a detailed analysis on the "usage" of a given domain ontology.

Chapter 5
Investigation Phase: Empirical Analysis of Domain Ontology Usage (EMP-AF)

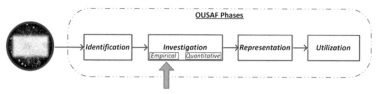

Focus of the Chapter

5.1 Introduction

[1]As highlighted in the previous chapters, in order to make effective and efficient use of an ontology, it is important to understand how a given ontology is being used by the users and its level of adoption. The next phase of the OUSAF, after the identification phase, is therefore the investigation phase, in which the analysis of the use of domain ontologies on the Web takes place. As mentioned in Sect. 1.7 there are different types of ontology user, such as Ontology Owner, Data Publisher and Application Developer, and each type of user requires different kinds of insight or information pertaining to the ontology usage, as briefly described below:

---

[1]Parts of this chapter have been:

1. Republished with permission of John Wiley & Sons from Empirical analysis of domain ontology usage on the Web: eCommerce domain in focus, Jamshaid Ashraf, Omar Khadeer Hussain, Farookh Khadeer Hussain, Volume 26, Issue 5, Copyright 2013 John Wiley & Sons, Ltd; permission conveyed through Copyright clearance centre.

2. Republished with permission of John Wiley & Sons from Making sense from Big RDF Data: OUSAF for measuring ontology usage, Jamshaid Ashraf, Omar Khadeer Hussain, Farookh Khadeer Hussain, Volume 45, Issue 8, Copyright 2014 John Wiley & Sons, Ltd; permission conveyed through Copyright clearance centre.

© Springer International Publishing AG 2018
J. Ashraf et al., *Measuring and Analysing the Use of Ontologies*, Studies in Computational Intelligence 767, https://doi.org/10.1007/978-3-319-75681-3_5

Ontology Owner: Ontology owners would be interested in knowing the following details:

- What is the adoption level of my ontology?
- Who is using it?
- Which specific components of the ontology are being used?

The answers to these questions will help an ontology owner to evaluate the usage of his/her ontology. The availability of this type of information provides a pragmatic feedback loop to the ontology evolution process, as shown in Fig. 1.10. Therefore, having such information is essential for ontologies to remain useful on the Semantic Web.

Data Publishers: Data publishers would be interested in knowing the following details, either for a given ontology or about their application area:

- Precisely what is being used by other data publishers from a given ontology?
- Which concepts of a given ontology are being used more and which concepts are being linked using which relationships?
- How is a domain-specific entity being attributively described?

The answers to these questions will help data publishers to understand what ontologies are being used for, and how, in their respective application areas. The availability of such information is necessary for data publishers to realize the benefits they will achieve by reusing existing ontologies. As mentioned in Sect. 1.6, by adopting (or reusing) used ontologies, a positive network effect, which means increasing the overall perceived utility of ontologies, is achieved. Furthermore, the increased use of an ontology by the community helps it to become the defacto structure (or schema) to represent the respective application area (or domain) [8].

Application Developer: To effectively and efficiently consume Semantic Web data (published on the Web), application developers need to know:

- What terminological knowledge of an ontology is available for use on the Web?
- Which concepts of a given ontology are being used more and how are these concepts being interlinked (using which relationships)?
- What are the common data and knowledge patterns available?
- How are entities being annotated or textually described?

The answers to these questions are important to the application developer because they provide a snapshot of the prevailing schema of the structured data published on the Web, allowing developers to program routines accordingly for the efficient and effective retrieval and consumption of semantically rich data. Knowing how entities are being described helps developers to query specific information about entities and to develop the operation and interfaces accordingly.

To obtain such an erudite insight into the use of ontologies from a range of perspectives for the different groups of users, a framework to analyse domain ontology usage is needed. The framework needs to be based on *real world instance data* to provide a practical insight from multiple perspectives and should *cover many aspects* to fulfil the needs of a wide range of users. There are two different ways to perform an analysis that is capable of providing the required information and insight on ontologies, as is explained in the next section.

5.2 Different Ways of Analysing Domain Ontologies

The use of ontologies to semantically describe the data on the Web has recently picked up pace to take advantage of the benefits offered by Semantic Web technologies. However, being in the early stage of adoption, there is limited understanding of how ontologies are actually being received by end users. For example, a data publisher makes use of a certain portion of the ontology and its components, based on his/her requirements, which could be different to the components used by other data publishers, depending on their requirements. The need to understand such usage patterns by different data publishers and other users is important, as explained in the previous section. In order to comprehensively understand how different users are using ontologies and the prominent and prevalent structures emerging from their usage, a *neutral observational approach* is required that provides an impartial empirical perspective on their usage. To then translate these neutral observations into actionable knowledge, a *quantitative measures approach* to ontology usage is required to determine the usage of domain ontologies. These two different but interlaced approaches to analysing ontology usage (see Fig. 5.1) provide a multi-view of the ontology usage landscape, and are briefly described in the next subsections.

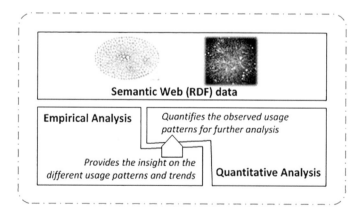

Fig. 5.1 Two different ways to analyse ontology usage

5.2.1 Empirical Analysis of Domain Ontology Usage

Empirical analysis is aimed at obtaining a neutral observation on the Semantic Web data to identify the prominent and prevalent structures emerging from the present use of domain ontologies. The need for empirical analysis at this stage of ontology adoption is rightly noted by Herman [143] (W3C Semantic Web Activity Lead):

> [...] we are at the point when we can measure what we got, and we
> can therefore come up with empirical data
> that will help us to concentrate on what is essential [...]

The empirical analysis of ontology usage on Semantic Web data, which essentially comprises schema-level and instance-level data, needs to analyse and extract patterns of ontology usage. While observing the schema-level data, which takes the form of terminological statements (T-Box) and instance-level data which takes the form of assertional statements (A-Box), it is important to decide on the *aspects* which need to be observed. In this case, aspects refers to the different viewpoints from which Semantic Web data has to be analysed. Each aspect offers a unique set of requirements necessitating different approaches and techniques to explore them. Terminological knowledge, which is encoded in the RDF statements by making use of the URI references defined by the domain ontologies, is considered during the empirical investigation to analyse the key aspects of ontology usage. The important aspects that are relevant to terminological knowledge analysis and helpful in addressing the requirements of different users (described in Sect. 5.1) are as follows:

1. **Understand how different vocabularies are interlinked at the instance level**: At the schema-level, this involves how different terminological statements originating from different ontological namespaces are being used to describe domain-specific entities. On the Semantic Web, the RDF data model and ontologies allow decentralized entities across different sources and domains to be linked. An understanding of how entities are linked at the schema level across various ontologies helps in extracting schema patterns and analysing entity linkage that exists within the dataset [203]. For ontology owners and application developers, it is useful to know the relationships present at the schema level to understand the users approach toward semantically describing the domain entity, as well as for preparing routines to query them.

2. **Understand how a concept is instantiated and described**: How are the pivotal concepts which represent the core elements of the domain used to describe the entities? To establish a thorough understanding of the use of pivotal concepts, it is important to know its instantiation, what other concepts contribute to its semantic description, what relationships it maintains with other concepts, and what attributes are used to provide factual knowledge.

3. **Understand the availability of textual description for human readability**: In order to allow a semantic application developer to consume the information distributed across remote systems and develop interfaces for human interpretation, knowledge regarding the use of textual description is important. Information about the presence of annotation and labelling properties enables application

developers to develop data-driven interfaces which are quite different from the classical form-based interfaces [72]. Ell et al. [89] listed a few of the benefits of labels, which include displaying human readable information instead of displaying URIs, using labels for indexing ([8] also highlighted similar benefits) and support for keyword and question-based searches over the web of data.

4. **Understand the data and knowledge patterns prevalent in the dataset**: Whether querying an anonymous dataset (triple store) whose schema is not known (unlike in traditional databases (RDBMS) where schema is known) or posing a federated query over the Semantic Web, it is very helpful and convenient to have some idea in advance about the nature of the data expected from the data source. For example, a prototypical query based on common patterns invariantly appearing across several data sources helps to generate a relaxed (generalized) query to start exploring the dataset. Therefore, it is helpful to have some understanding about the knowledge and data patterns available in the dataset to generate prototypical queries.

To empirically understand the use of domain ontologies in relation to these aspects, the **EMP**irical **A**nalysis **F**ramework (EMP-AF) is proposed in this chapter.

5.2.2 Quantitative Analysis of Domain Ontology Usage

While the above empirical analysis provides an overview of ontology usage from a neutral perspective to understand the use of domain ontologies, taking these impartial observations into actionable knowledge requires quantification of the observation. In other words, empirical analysis identifies the *key factors* involved in proliferating and driving ontology adoptions, but to utilize the key factors so that they can be used in various scenarios such as ranking, indexing and querying the information, they need to be quantified. These key factors lead to the development of more focused metrics to measure ontology usage by considering the conceptualized model represented through the ontology. To undertake quantified analysis of ontology usage on the Web, the **QUA**-ntitatible **A**nalysis **F**ramework (QUA-AF) is proposed in the next chapter.

The remaining sections of this chapter are organized as follows. Section 5.3 presents the EMP-AF and its two phases, namely data collection and aspect analysis. Section 5.4 defines the metrics used to empirically analyse the use of domain ontologies. To explain the working of the EMP-AF, a case study is described in Sect. 5.5 which will be used as an example to analyse domain ontology usage. Section 5.6 discusses the implementation of the data collection phase and details the dataset characteristics of the case study. Section 5.7 provides details on the results obtained by analysing domain ontology usage, based on the metrics developed as part of EMP-AF. Section 5.8 presents a discussion on the analysis of the EMP-AF by considering the requirements of different types of user, as discussed in Sect. 5.1. Finally, Sect. 5.9 concludes the chapter.

5.3 EMPirical Analysis Framework (EMP-AF)

The **EMPirical Analysis Framework (EMP-AF)** comprises two phases, namely the *data collection phase* and the *aspects analysis phase*, as shown in Fig. 5.2. The data collection phase is responsible for collecting the real world instance data necessary for empirical analysis, whereas the aspect analysis phase is responsible for analysing the use of domain ontologies from multiple aspects to obtain the insight required by users, as mentioned in Sect. 5.1. In the next subsections, the objectives and working details of each phase are presented.

5.3.1 Data Collection Phase

To obtain erudite insight into the use of ontologies and their components in a real world setting, it is of paramount importance that data are collected from the data sources that are using domain ontologies to describe their data. The identification phase of the OUSAF (Chap. 4) provides the candidate ontologies which are being used by data publishers in a given application area. Identifying these ontologies helps to find potential data sources which use these ontologies, and they can be included in the data collection phase of the EMP-AF.

The data collection process (see Fig. 5.2a) crawls the Web to collect the Semantic Web data published by data publishers. This means that the crawler responsible for collecting the data needs to be aware of the various ways in which structured data is published on the Web and the different serialization formats that are used on the Web to publish Semantic Web data. Aside from the infrastructure requirements, the crawling process should be able to deal with network issues which may arise during the crawling process.

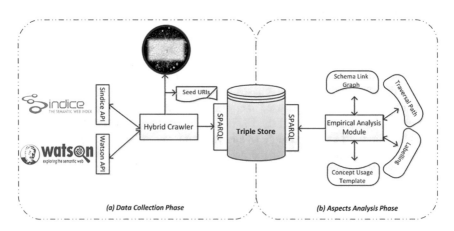

Fig. 5.2 Empirical analysis framework (EMP-AF)

To address the abovementioned requirements and gather real world instance data described using ontologies, a hybrid crawler is proposed as part of the EMP-AF. The details of the hybrid crawler and the collected dataset are described in Sect. 5.6.

5.3.2 Aspects Analysis Phase

The aspect analysis phase (see Fig. 5.2b) of the EMP-AF focuses on the execution of empirical analysis. This phase comprises the Empirical Analysis Module which implements four metrics to investigate the dataset from multiple aspects and a data access component to evaluate SPARQL queries. The four metrics implemented as part of the EMP-AF are introduced below.

1. To understand how different vocabularies are interlinked at the instance level, the first metric we consider is the **Schema Link Graph (SLG)** which reveals the relationship between vocabularies at the instance level, based on the use of the terminological statements of ontologies in the dataset. Hence, SLG addresses the first requirement of empirical analysis by helping ontology owners and application developers to understand the semantic relationships present on the Web in the given application area, and to use these for subsequent processes.
2. To understand how a concept is instantiated and described to obtain a detailed usage analysis, the **Concept Usage Template (CUT)** is proposed. This captures the instantiation of concepts, the relationships of the concept, and the data properties used to describe it. It also captures the different vocabularies being co-used with this concept. This detailed multi-perspective insight provided by CUT helps all types of ontology users to glean relevant information.
3. To understand the availability of textual description for human readability, the **labelling** aspect is proposed. It captures the use of properties for labelling purposes. Labelling benefits application developers by helping them to better understand the available textual descriptions, as mentioned in the third requirement of empirical analysis in Sect. 5.2.1. As good practice, data sources make use of labelling properties which are either part of the standard vocabularies or popular in the community; therefore, in formulating the labelling properties, one needs to consider all these different usage patterns.
4. To understand the data and knowledge patterns prevalent in the dataset, the **Traversal path** structure is constructed to capture the prevalent knowledge patterns in the domain ontology usage and understand the invariant patterns available to assist in accessing information. Traversal paths extract the knowledge and data patterns available in the dataset to facilitate the generation of prototypical queries, as mentioned in the fourth requirement of empirical analysis. Travelling the graph, especially an RDF graph, which is a multi-edge and directed graph, is a computationally expensive operation, therefore to find the occurrence of different patterns, it is necessary to consider a preprocessing stage to reduce the overall computation time.

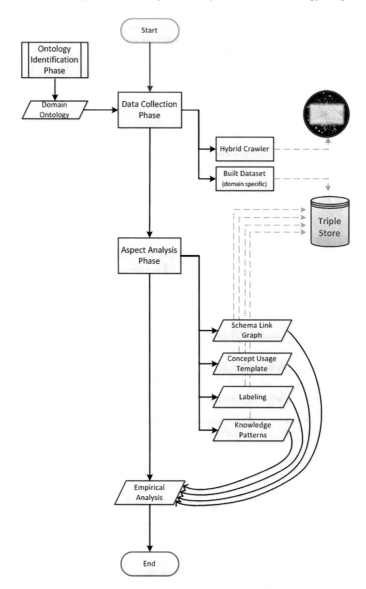

Fig. 5.3 Flow of activities in EMP-AF

These metrics help to address the requirements of different users (discussed in the introduction to this chapter) and the aspects highlighted in Sect. 5.2.1. Section 5.4 formally describes these metrics.

In the next subsection, the set of sequential activities carried out as part of the EMP-AF is presented.

5.3.3 Sequence of EMP-AF Activities

The EMP-AF comprises two phases, namely the data collection phase and aspect analysis phase. Each phase involves a certain number of activities to carry out the required functionality. In order to provide an overview of the flow of tactivities and their sequence, a summary is presented in Fig. 5.3.

- In the data collection phase, a dataset relevant to an application area (domain focused) is collected.

  - The hybrid crawler is implemented to crawl the relevant Semantic Web data.
  - The crawled data is populated into the triple store.

- The data is analysed using the metrics defined in the aspect analysis phase.

  - To reduce the computation cost of the resource intensive operation, pre-processing is carried out.
  - Using SLG, the relationships present in the dataset are analysed.
  - Using CUT, the use of pivotal concepts are analysed.
  - The labelling present in the dataset is observed.
  - The knowledge patterns are observed by constructing the traversal paths.

- The results are analysed to infer the use of domain ontologies on the Web.

5.4 Metrics for EMP-AF

In this section, the metrics used for empirical analysis as part of the EMP-AF are presented. Additionally, to explain the analysis obtained from each metric, a sample RDF Graph (Fig. 5.4a for SLG and Fig. 5.5 for other metrics) is used which provides an overview of the computation process and the results obtained from each metric.

Before proceeding with discussion on the metrics, the necessary preliminaries for the metrics are defined.

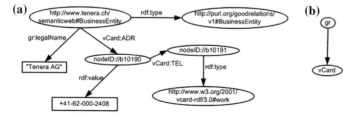

Fig. 5.4 a Sample RDF graph from the dataset with blank nodes and **b** the corresponding Schema Link Graph

```
 1  @base <http://www.example.com/websource#>
 2  @prefic gr:<http://purl.org/goodrelations/v1#>
 3  @prefix dc:<http://purl.org/dc/terms/>
 4  @prefix vso:<http://purl.org/vso/ns#>
 5  @prefix coo:<http://purl.org/coo/ns#>
 6  @prefix foaf:<http://xmlns.com/foaf/0.1/>
 7  @prefix vCard:<http://www.w3.org/2001/vcard-rdf/3.0#>
 8  :cardealer
 9      rdf:type gr:BusinessEntity ;
10      dc:title "business entity and car data";
11      dc:date "2009-09-12";
12      gr:legalName "The Example Company";
13      owl:sameAs  <http://www.acme.com/example>;
14      foaf:homepage <http://www.example.com>;
15      rdfs:seeAlso <http://www.example.com/about.pdf>;
16      gr:offers ex:Offering_1.
17  ex:Offering_1
18      rdf:type gr:Offering;
19      gr:includes  ex:product_1;
20      rdfs:comment "Eco Car on sale"@en;
21      gr:availableAtOrFrom    ex:location;
22      gr:category "Used Car".
23  ex:product_1
24      rdf:type gr:ProductOrServiceModel;
25      rdf:type vso:Automobile;
26      rdf:type coo:Derivative;
27      gr:name   "The Blue Car";
28      gr:category "Automobile";
29      gr:color "Red".
30  ex:location
31      rdf:type gr:LocationOfSalesOrServiceProvisioning;
32      vCard:ADR [
33        vCard:Street   "2253 Jackson Ave.";
34        vCard:Pcode    "00553";
35        vCard:City     "New York";
36        vCard:Country    "US".
37      ].
```

Fig. 5.5 Sample RDF code for discussion

5.4.1 Preliminaries

RDF Triple (triple): A $triplet := (s, p, o) \in (U \cup B)\ X\ U X (U \cup B \cup L)$ is called
an RDF triple, where s is called subject, p predicate, and o object.

Class: Class is referred as an $RDFTerm$ which appears in either

- o of a triple t where p is rdf:type; or
- s of a triple t where p is rdf:type and o is rdfs:Class or owl:Class

Property: Property is an $RDFTerm$ which appears in either

- p of a triple t; or
- s of a triple t where p is rdf:type and o is rdf:Property

Instance of a Concept (C): A triple $t = (s, p, o)$ or set of triples in the dataset is an
instance of a triple pattern $t_c = (s_c, p_c, o_c)$ if there exist

- s_c is URI Reference
- p_c is `rdf:type`
- o_c is the *class* (Concept) of domain ontology.

In the next subsection, the metrics used in the aspect analysis phase to empirically
analyse ontology usage are presented. Additionally, the analysis obtained from each
metric is explained with the help of an RDF graph.

5.4.2 Schema Link Graph (SLG)

The Schema Link Graph (SLG) is an undirected graph consisting of a finite set of
vertices V and a set of edges E, representing a link between the two vertices. SLG
is used to study the relationship between different ontologies in describing entities.
Formally, the Schema Link Graph is defined as follows:

Schema Link Graph (SLG): *The Schema Link Graph (SLG) is a tuple (V, E),
where n is a node (n $\in$ V) such that n is the ontology namespace used in the
dataset. 'Used' means the presence of a triple where n appears as an object (for
instantiation with* `rdf:type`*), or in a predicate to describe the object. E is the
edge set and e $\in$ E is an edge of graph V linking two nodes n_1 and n_2 such that
either there is a triple which entails that n_1 is the namespace of the subject and n_2 is
the entailed namespace of the object, or there is a m sequence of triples connected
through a blank node such that n_1 is the entailed namespace of the subject of the
first triple and n_2 is the namespace of the object in the mth triple where m > 1.*

Example: To give an example, Fig. 5.4a shows an RDF graph snippet extracted from
the RDF graph representing the semantic data published by http://www.tenera.ch.
Here, the sequence of triples (subject, predicate, object) is connected to semantically describe the entities (resources) using ontologies. In the sample RDF graph,
there are schema level triples creating individuals of type class defined in the domain
ontology and the instance level triples describe the entities. For the construction
of SLG, triples with the rdf:type predicates are retrieved and if there is a relationship joining the resources instantiated using different namespaces (ontologies), a link
between these two ontologies is created. As shown in Fig. 5.4a http://www.tenera.ch/
semanticweb#BusinessEntity, a resource is defined by the GRO concept and linked
using cVard:ADR with a resource of vCard vocabulary. Therefore, in the resultant
SLG, there are two nodes *gr* and *vCard* with an edge. Figure 5.4b shows that there
is a URI in the RDF graph of type *gr* directly or indirectly (through blank nodes)
connected with the URI of type *v*.

5.4.3   Concept Usage Template (CUT)

The Concept Usage Template (CUT) captures how a concept is used in the dataset and what properties (both domain ontology predicates and other predicates) are used to describe the entities instantiated by the concept. The template attempts to capture the ubiquitous patterns available and arranges them to facilitate the processing of information for specific purposes, such as searching, browsing, querying and reasoning.

The template captures six aspects of concept usage. It looks at the RDF graphs available in the dataset and analyses the concepts instantiation, the use of different vocabularies in describing entities, the presence of different relationships, the use of different data properties, the use of other concepts to describe the same entity and the presence of other constructs to provide additional and supplementary information about the entities represented by the concept.

CUT comprises the following metrics:

5.4.3.1   Concept Instantiation

This refers to the number of instances instantiated by the class representing the concept. This gives us the number of entities available in the dataset and reflects the dominance of the entity in the dataset when compared with the templates of other concepts. In most Web ontologies, subsumption axioms are used to provide the taxonomical relationship between concepts, and with inference, provisioning the concept instantiation may fluctuate depending on where the concept falls in the taxonomy hierarchy. Since most triple stores implement RDFS entailment rules, it is safe to consider the $rdfs9^{2}$ rule while measuring the instantiation of a top level concept in the taxonomic hierarchy. The *concept instantiation (CI)* of a concept C is given as follows:

$$CI(C) = |triples|where, \begin{cases} s = RDFTerm \\ p = rdf : type \\ o = class\ defined\ by\ ontology \end{cases} \tag{5.1}$$

In the case of subsumption axioms [114], the $CI(C)$ can include the instances instantiated by the sub-concepts (subclasses) of o such that:

$$o = entail_{rdf9}(C) \tag{5.2}$$

where $entail_{rdf9}(C)$ is a function which implements the $RDFS9$ rule:

```
IF (uuu rdfs:subClassOf xxx AND vvv rdf:type uuu)
   THEN (vvv rdf:type xxx)
```

---

[2]IF(`<v subClassOf w>` and `<u type v>`) THEN `< u type w>`.

$CI(C)$ returns the numeric value representing the number of entities defined by the concept and its sub-concepts.

5.4.3.2 Vocabs

Vocabs provides the list of ontologies (other than the domain ontology) used to describe the entity. Ontologies are represented here with their namespace prefixes and include both the predicates ontology prefix and the prefix of the concept to which it is linked. Vocabs help in understanding the different ontologies which are co-used to describe different aspects of the entity. Formally, $Vocabs$ is defined as:

Definition: $Vocabs$ is a set of namespaces (empty possible) of the vocabularies used in a *triple* such that o is the domain ontology concept and p is the URI reference of the ontology other than the domain ontology used to describe the s.

$$Vocabs = \{vocab_1, vocab_2 \ldots vocab_n\} \tag{5.3}$$

such that $vocab_i$ is the namespace of the $p's$ URI reference.

5.4.3.3 Object Property Usage

This provides a list of relationships available to describe the entity by relating it to other sets of entities and resources. It includes the properties defined by the domain ontology as well the properties of the ontologies listed in $Vocabs$. Object property usage allows an understanding of the available information pertaining to the entity and its richness by exploring the entities linked to it through these properties.

$$Object\,Pro(C) = \{pre_1, pre_2 \ldots pre_n\} Such\,that\,pre_i = Property \tag{5.4}$$

The $Object\,Pro(C)$ set contains the *URI references* representing the object properties defined by the ontologies belonging to $Vocabs$.

5.4.3.4 Attribute Usage

This provides the textual information about the entity. This may include the RDF label properties and the data type properties of the domain ontology and non-domain ontologies. A textual description linked with entity instance is useful information for data processing and the user interface.

$$Attri(C) = \{att_1, att_2 \ldots att_n\} \tag{5.5}$$

such that $att_i \in (L_p \cup L_t)$.

The *Attri(C)* set contains the *URI references* representing the datatype properties defined by the ontologies belonging to *Vocabs*.

5.4.3.5   Class Usage

Class usage records the list of other concepts of which the entity is a member. This allows more to be learned about the entity, as different concepts, when used to instantiate the same entity and define the broader view, reflect the reality being represented by the entity. It is believed that class usage provides the conceptual overlap which exists between related but different concepts formalized by different ontologies, and that it can be exploited to generate semantic mapping between related terms.

$$ClassUsage(C) \text{is set of classes such that there exists a triple in} \qquad (5.6)$$

the dataset where $p = rdf : type$ and o is **class** and $o \neq C$

5.4.3.6   Interlinking

Interlinking provides a list of linking properties used to create links across datasets. Examples of such links are link base and equivalence link [85]. Here, the main focus is on equivalent links, which helps to specify the URIs that refer to the same entity or resource. Semantic Web languages provide built-in support for creating equivalent links between different components of the ontology and data. The resources and entities are linked through the `owl:sameAs` relation which tells the applications that these two resources (subject URI and object URI) are describing the same entity, and that their data can be merged to obtain an exploded view of the entity. Thus, interlinking is obtained by identifying the use of any interlinking property for a given entity.

Example: The above-mentioned analysis methods and metrics are explained using a sample RDF graph. Figure 5.5 shows the sample RDF graph of a fictitious "Example.com" data source. The RDF data describes a company which is in the car sales business. The triples in the RDF graph represent information regarding the business entity, its shop/office location (address), and the offers and products included in the deal. For the sake of brevity and readability, relevant triples in turtle syntax are listed and will be used in this section for discussion and explanation. Lines 1–7 of the sample RDF code contains the prefixes which are used in the triples to access the vocabulary (or terms) defined by their respective namespaces to describe the resources (entities). Lines 8–37 of the sample RDF code describe the different resources linked through relationships to semantically describe the entities.

The CUT of the entity (i.e. ex:cardealer) of type `gr:BusinessEntity` is shown in Table 5.1. In the sample RDF graph, `ex:cardealer` is the business

Table 5.1 Sample RDF
Graph: CUT of
gr:BusinessEntity
(ex:cardealer)

Entity	gr:BusinessEntity
Instantiation	1
Vocabs	gr, dc, foaf
Object properties	gr:offering
Attributes usage	de:title, de:date, foaf:homepage Class Usage
Interlinking	rdfs:seeAlso

entity which is an instance of the type `gr:BusinessEntity` class defined in the GoodRelations ontology. The value of the concept instantiation using Eq. 5.1 is $CI(C) = 1$, as there is only one instance of the type `gr:BusinessEntity`. *Vocabs* is the set of prefixes used to describe the entity and in this example, Eq. 5.3 returns $Vocabs = \{gr, dc, foaf\}$. Note that in *Vocabs*, the prefixes of W3C-based standard languages such as RDF, RDFS and OWL are not considered to be the focus, which is more on the domain ontologies. In the case of object property usage, Eq. 5.4 returns $ObjectPro(C) = \{gr : offering\}$ and for attribute usage, Eq. 5.5 returns $Attr_i(C) = \{dc : title, dc : date, foaf : homepage\}$. The class usage of the product entity (i.e. ex:product_1, line 23) of type `gr:ProductOrServiceModel` returns the set of classes of which the entity is also a member, i.e. $ClassUsage(C) = \{vso : Automobile, coo : Derivative\}$ (not shown in Table 5.1 which covers the CUT of gr:BusinessType. Additionally, link base and equivalence links are provided to allow users to access additional relevant information (`rdfs:seeAlso`; line 15) and explode the information about the entity by merging the description published on two different locations (`owl:sameAs`; line 13), i.e. $Interlink(C) = \{rdfs : seeAlso, owl : sameAs\}$.

5.4.4 Labelling

Labels are the textual information provided with the entity description to allow a better understanding of the entities before these entities are processed by Semantic Web applications. The emphasis is on analysing how labelling properties are used with entity description, which is helpful for information retrieval and presentation.

While analysing the entity, we look at the use of various label properties in the data and discuss their usefulness in scenarios such as finding hidden information from the label text, using language tags to facilitate the internationalization of semantic applications, and developing the user interface for information which is syntactically published for machine consumption.

5.4.4.1 Formal Labels

RDFS specification provides two properties, `rdfs:label` and `rdfs:comment`, to provide human-readable information about the resources. The former is normally used to provide a human-friendly version of the resource name, which is otherwise an opaque URI, and the latter is used to present a human readable description of the resource. These two label properties are referred to as **formal label (fl)** when analysing the presence of label properties in the dataset in general and in the entity description specifically. Such online documentation on resources is very useful and domain ontologies often define more specific labelling properties.

The following metric is defined to measure the use of *fl* for each pivotal entity. $Entity_{fl}$ measures the ratio of entities with at least one formal label to all pivotal entities in the dataset.

If C is the concept of the domain ontology (class) then:

$$fl = \{rdfs : label, rdfs : comment\}$$

$$Entity_{fl}(C) = \text{ number of instances}(C) \text{ with fl/total number of instances } (C)$$
$$(5.7)$$

5.4.4.2 Domain Labels

There are two common practices for defining domain ontology label properties: first, by describing label properties as the subproperty of `rdfs:label` using the subproperty axiom (subsumption), and second, by having a datatype property with `rdfs:Literal` as its range. In some cases, the label properties are defined by specifying literal datatype and in such cases, `xsd:string` datatype is used. Here, these domain-ontology-defined label properties are referred to as **domain labels (dl)**. Ell et al. [89] have proposed label-related metrics to measure the completeness, the efficient accessibility of label properties, and the unambiguity of the labels in the knowledge base. These metrics help to quantify the presence of labels in a dataset, however to understand their usefulness in a real setting for information retrieval and presentation purposes, one needs to analyse label properties for each pivotal entity and discuss their usefulness.

Likewise, $Entity_{dl}$ computes the ratio of entities with at least one domain ontology label to all pivotal entities in the dataset. The sum of these two measures tells us how rich a particular concept (pivotal entity) is in terms of labels. If C is the concept of the domain ontology (class) then:

$$dl = \{ i| i \text{ is the label property defined in domainontology}\}$$

$$Entity_{dl}(C) = \text{number of instances}(C) \text{ with dl/total number of instances } (C)$$
$$(5.8)$$

Example: To use an example to explain what labels are available and how they are used in the knowledge base by using metrics, namely $Entity_{fl}$ and $Entity_{dl}$, let us refer back to the sample RDF graph (see Fig. 5.5). The focus is on gr:BusinessEntity as the pivotal concept, the label metrics for the entity of type gr:BusinessEntity is measured using Eqs. 5.7 and 5.8, respectively.

$$Entity_{fl} = 0/1 = 0$$

$$Entity_{dl} = 1/1 = 1$$

The label attributes used for the description of the ex:cardealer entity are listed from lines 9–16 of the sample code. For $Entity_{fl}$, only RDFS-based label properties (i.e. rdfs:label and rdfs:comments) are considered and none of them is used in this particular example. There is only one instance of entity (individual) of type gr:BusinessEntity therefore $Entity_{fl}$ equals zero. Likewise, for $Entity_{dl}$, gr:legalName predicate usage, which is a domain ontology label property (the complete list of domain ontology labels are discussed in Sect. 6.5), is present, therefore the value of $Entity_{dl}$ is 1.

5.4.5 Knowledge Patterns (Traversal Path)

A traversal path determines the sequence in which properties are used to access the description of related entities within a given context. A traversal path starts with the instance of the entity class in focus and follows the available sequence of instance-property-instance triples to record all the paths in the dataset. The following metrics pertaining to traversal paths are defined.

5.4.5.1 Unique Paths

Unique paths computes the number of unique paths leading from the entity (out links). One entity can have zero or many paths of varying lengths, depending on the RDF graph in the dataset. A complete set of unique paths helps in understanding the data patterns available, which can further assist in querying the dataset.

5.4.5.2 Average Path Length

Average path length helps in understanding the entity description depth available in the dataset.

5.4.5.3   Max Path Length

Max path length helps in understanding the maximum possible description depth available in the knowledge base.

5.4.5.4   Path Steps

Path steps helps in identifying the triples found in the traversal paths.

In traversal paths, unique paths available in the RDF graph (or dataset) and the maximum and average traversal path lengths are computed. The traversal path procedure constructs the list of all available paths in the dataset and this list of paths is then used to compute the maximum and average path length. Additionally, the path steps of each path are generated and their frequency in the path list is computed to reflect the occurrences of each path step in the path list. As mentioned earlier, the computation of these metrics on a large graph becomes computationally expensive, therefore preprocessing is done on the dataset to make the computation process practical.

Example: In the example code, there are two unique paths in the RDF graph, one of length 3 and the other of length 2 (see Fig. 5.6). The length is computed by counting the number of predicates (relationships) available in a path. The path steps and their strength value are shown in Fig. 5.7. It can be seen that the first path step has a strength of 2 because this appears in two paths, and the remaining path step only has a strength value of 1 because this appears once in both paths. Paths and path steps provide a snapshot of the knowledge in the form of triple patterns that indicate

Fig. 5.6 Traversal paths

Path step	Strength
gr:Business Entity — gr:offers → gr:Offer	2
gr:Offer — gr:availableAt orFrom → gr:LocationOfSalesOr ServiceProvisioning	1
gr:LocationOfSalesOr ServiceProvisioning — vCard:adr → [ ]vCard:ADR	1
gr:Offer — gr:includes → gr:ProductOr ServiceModel	1

Fig. 5.7 Path steps and their strength

the invariance of instance data or entity description across the data sources that are contextually relevant (domain specific).

In the next section, a case study is presented to introduce the domain ontology on which the analysis will be performed.

5.5 Case Study: Empirically Analysing Domain Ontology Usage

One of the domain ontologies identified in the identification phase of the OUSAF is the GoodRelations Ontology (GRO). GRO, its schema and key concepts of the ontology are described to introduce the conceptual model represented by the ontology. This ontology will be used in the subsequent section to empirically analyse the use of domain ontologies on the Web.

5.5.1 GoodRelations as a Domain Ontology

GoodRelations [142] is one of the first Web ontologies of its kind, developed and introduced in 2008 to conceptualize the eCommerce domain on the Web. From the outset, GRO has allowed businesses to describe their company (Business Entity), offers and product-related data, based on the RDF data model, over the Web, which can be accessed and processed by Semantic Web applications and search engines. It has recently seen an increase in popularity and adoption (See Figs. 5.8 and 5.9) by the Semantic Web community, particularly after being recognized by major search engines such as Google (www.google.com), Yahoo (www.yahoo.com) and Bing (www.bing.com). GRO has been successful in selling the idea and value of explicit semantics to these search engines, which have, for a long time, processed unstructured data to extract fuzzy semantics algorithmically from documents.

5.5.2 Conceptual Schema and Pivotal Concepts

GRO is a kind of live ontology which is evolving with time to capture the changes and improve its conceptual representation of the domain model. The latest version of the GRO ontology comprises 31 concepts (classes), 50 object properties, 44 data properties and 48 named individuals. Keeping backward compatibility intact, the ontology model is updated frequently to add new object and data properties, based on the experience and feedback gained through real world implementations. The GR model[3] has three main concepts. They are the *Business Entity*, *Offering* and *Product*

---

[3]http://www.heppnetz.de/ontologies/goodrelations/goodrelations-UML.png; retr.; 27/11/2017.

Prominent Users of GoodRelations

GoodRelations is being used by 10,000+ small and large shops world-wide. On this page, we list very prominent users.

Current Users

- **Google** officially recommends GoodRelations for sending structured information for Google Rich Snippets to Google (since 11/2010).
- **Yahoo** officially recommends GoodRelations for sending structured information for their SearchMonkey feature (since 10/2008).
- **Best Buy** is using GoodRelations as fundamental part of their digital marketing strategy and publishes full catalog, store, and special offer with GoodRelations on their production Web sites.
- **O'Reilly** is using GoodRelations for Semantic SEO of all of their book titles.
- **Volkswagen UK** is using GoodRelations for exposing car feature and car component information at massive scale.
- **Renault UK** is using GoodRelations for Semantic SEO for their merchandise shop.
- **OpenLink Software** is using GoodRelations as the fundamental vocabulary for E-Commerce technology based on Virtuoso and other products.
- **Peek & Cloppenburg** is using GoodRelations for publishing information on all European stores plus the brands available in each one of them.
- **CSN Stores** is using GoodRelations for Semantic SEO of all of their 2,000,000 item pages and substores.
- **Arzneimittel.de**, one of Germany's leading mail order pharmacies, is using GoodRelations in RDFa on all of their ca. 250,000 item pages.

Fig. 5.8 GoodRelations Ontology Adopters (http://wiki.goodrelations-vocabulary.org/References; retr., 26/11/2017)

Ping the Semantic Web

These namespaces are used to describe entities in X number of documents

Namespaces *(2016 know namespaces)*	Number of documents
http://xmlns.com/foaf/0.1/	1, 406, 142
http://www.w3.org/2002/07/owl#	671, 096
http://purl.org/goodrelations/v1#	651, 045
http://blogs.yandex.ru/schema/foaf/	585, 037
http://sites.wiwiss.fu-berlin.de/suhl/bi...	252, 373
http://rdfs.org/sioc/ns#	222, 140
http://www.w3.org/2003/01/geo/wgs84_pos#	181, 821
http://rdfs.org/sioc/types#	142, 558
http://purl.org/ontology/bibo/	129, 527

Fig. 5.9 GoodRelations instances on the web of data

or Service. Each concept focusses on a specific aspect of the eCommerce domain. GRO is available at http://purl.org/goodrelations/v1 and **gr** is the prefix used in this chapter and elsewhere to refer to the vocabulary namespace defined by GRO.

5.5.2.1 Business Entity

The `gr:BusinessEntity` concept represents a business organization (or individual) which intends to offer or seek products on the Web. The main purpose of this concept is to provide the attributes needed to describe any business, such as the name of the company, address, location, vertical industry in which it operates and any other identifier which makes it uniquely distinguishable on the Web. None of the above-mentioned properties are mandatory for describing the business entity (company or individual) using GRO; however the more information that is available, the easier it is to find and consume information with high precision. For large organizations that have multiple outlets or shop locations, GRO provides concepts (`gr:Location` and deprecated `gr:LocationOfSalesOrServiceProvisioning`) to describe shops or service centres through which products or services are provided. Each shop location has its own operating hours which are described using the opening hour specification (`gr:OpeningHoursSpecification`).

5.5.2.2 Offering

`gr:Offering` is the pivotal concept in the GRO. This concept allows the description of a particular offering a business entity is likely to make or seek on the Web. The version on the ontology we considered, there are 15 data type properties (all optional) to describe offer details such as availability, validity, name and description of the offering. Name and description are recent additions which allow users to learn more about the offer. Offering can include one or more products with a price specification describable in any currency. It is possible to attach supplementary details such as warranty promises, customer eligibility for the offer, shipment options and charges, and acceptable methods of payment.

5.5.2.3 Product or Service

The third main concept is Product or Service (`gr:ProductOrService`). As mentioned earlier, an offering can contain one or more products (or services) and is usually described using one of the three possible subclasses of this main (abstract) class. GROs principal focus is to cover the conceptual model of offering rather than being a product ontology. However, `gr:ProductOrService` and its sub-concepts can be used to describe a product and its qualitative and quantitative properties to describe lightweight product ontology.

A description of the implementation of the data collection phase is presented in the next section.

5.6   Data Collection: Hybrid Crawler and Dataset

To gain a clear understanding of the RDF data and the use of ontologies to provide a shared inference and structure on the Web, a dataset comprising domain-specific data extracted from the Web is built to conduct an investigation on empirical grounding. In the real world, the interest is in data sources which use the domain ontology using core concepts to provide schema level metadata. In the following subsections, the approach adopted to identify potential data sources and the minimum selection criteria used is discussed. Then, the dataset collection approach, including hybrid crawling and the selection of seed URIs, followed by the dataset characteristics, is described.

5.6.1   Hybrid Crawler

One of the potential sources for the required data is the LOD cloud[4] which in the dataset we considered hosts 295 datasets containing approximately 32 billion triples in total. This appears to be a very fertile source of data for our study; however, as reported in [30, 145], the datasets in the LOD cloud are publishing more data and merely using ontologies, hence neglecting, if not failing, to provide schema level meta-information deemed necessary for information apportioning over the Web. The published LOD statistics also mention that 64.75% of the datasets have made use of non-W3C base-vocabularies (RDF, RDF Schema and OWL) which are here called open ontologies/vocabularies. Of these open ontologies, 78.31% of datasets mutually use the DC (Dublin Core) (31.19%), FOAF (27.46%) and Simple Knowledge Organization System (SKOS) (19.66%) ontologies to provide schema level information. *Noticeably, only 4 (1.36%) out of 295 are reported to have used GRO, yet on other hand,* www.PingTheSemanticWeb.com *ranks GRO as the third most used ontology after FOAF and OWL* (see Fig. 5.9). These numeric facts highlight the paucity of use and availability of ontological knowledge in the LOD dataset. Therefore, a dataset was built to collect the RDF data currently published using the domain ontology.

To collate domain-focused data, the minimum criteria employed for the selection of potential data sources is to identify the data publishers which have at least described the key concepts using the domain ontology. In our case, Business Entity and Offering are the primary identification drivers. A list of seed URIs for crawling using Sindice API[5] and the Watson[6] semantic search engine (see Fig. 5.10) was built. An initial attempt was made to use the semantic crawlers available, such as LDSpider [160], but since most of the eCommerce-related RDF data is embedded in HTML pages using RDFa, and because of a lack of interlinking between different resources even

---

[4]http://www4.wiwiss.fu-berlin.de/lodcloud/state/ (retr., 27/11/2017).

[5]As of Dec 2015 trading as https://siren.solutions (previously known as http://sindice.com/developers/api) (retr., 27/11/2017).

[6]http://watson.kmi.open.ac.uk/WatsonWUI (retr., 22/11/2017).

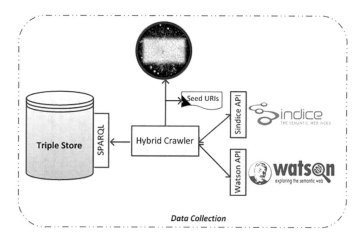

Fig. 5.10 Schemata diagram of Hybrid Crawler

within the same hostname, the crawler in their present implementation do not offer an effective solution. Therefore, a hybrid crawler which crawls in a similar way to traditional Web crawlers by following hyperlinks and extracting only the RDF triples available in Web documents, was implemented by extending LDSpider. Using REST-based Web services, namely Any23[7] and RDFa Distiller,[8] the extracted RDFa snippets from web documents were transformed into an RDF/XML document to produce one RDF graph for each Web document.

The RDF Graphs were loaded into the OpenLinks Virtuosos[9] triple store to create the dataset for further analysis known as **GRDS**. From an RDF data management perspective, named graphs [53] were used to group all the triples from one data source (hostname) under a uniquely-named graph International Resource Identifier[10] (IRI), allowing the dataset to be queried vertically (one data source) and horizontally (across data sources).

5.6.2 Dataset Characteristic

The empirical analysis was performed on the GRDS dataset which was built using the hybrid crawler discussed earlier. The GRDS dataset comprises 22.3 million triples (loaded into the open source version of the Virtuoso triple store) collected from 211 different data sources (pay-level domains). The complete list of data sources included in GRDS are shown in Fig. 5.11.

---

[7]http://any23.apache.org (retr., 27/11/2017).

[8]https://www.w3.org/2007/08/pyRdfa/ (retr., 25/10/2017).

[9]http://virtuoso.openlinksw.com/ (retr., 25/10/2017).

[10]https://www.w3.org/International/O-URL-and-ident.html (retr., 23/10/2017).

Data Sources	Data Sources	Data Sources	Data Sources	Data Source
www.3kbo.com	www.corvettespecialty.com	www.lilumtechnologies.com	www.openlinksw.com	www.stanwoo
www.abe.pc.pl	www.c-paintings.com	www.importsvipbolsas.com	www.opvallendeplanten.nl	www.starline.d
www.acigroup.co.uk	www.credis.org	www.inhead.pl	www.oreilly.com	www.store.ins
www.agushop.com	www.csnceilingfans.com	www.inndata.at	www.overstock.com	www.store-go
www.akw-fitness.de	www.csnstores.co.uk	www.internethq.com.au	www.palimpsest-press.com	www.store-ste
www.alibi.com	www.csnstores-com.com	www.internethq.com.au-swebarf	www.pano-store.com	www.succentu
www.allkitchencarts.com	www.customized.de	www.intisgifitalpacas.com	www.pauladeenstore.com	www.suitcase.
www.allmodern.com	www.cyelite.com	www.it-bestshopping.com	www.pdagroup.net	www.superska
www.anonbrand.com	www.diaper-dash.com	www.i-views.de	www.pearlsandpieces.dk	www.svanvit.s
www.arizona-realestate-market.com	www.discountcoffee.ie	www.jarltech.de	www.peek-cloppenburg.de	www.swimkitte
www.armazem4x4.com.br	www.discountofficehomefurniture.com	www.jing-shop.com	www.petsbycsn.com	www.swisstgal
www.asienraum.com	www.doctormaryjayne.com	www.joyfay.com	www.pharma2phork.org	www.symbolor
www.atacado-vipfolheados.com.br	www.dowclpnie.com	www.justvanities.com	www.pharma2phork.org	www.synapse-
www.atlanticlinux.ie	www.econoclick.com.br	www.karneval-alarm.de	www.piccadillys.com	www.tanzanite
www.bathroomfurnituredirect.com	www.erlebnisfuehrer.ch	www.karniyarik.com	www.pinpoint.benefits.com.au	www.technicin
www.beddingsets.com	www.erospa-shop.de	www.kasztany.com	www.plushbeautybar.com	www.tenera.ct
www.BestBuy.com	www.espacelibido.com	www.kedalsouvenir.com	www.praxis-kohn.de	www.thegreati
www.BestBuy.com Store	www.ews-ingenieure.com	www.kica-jugendstil.com	www.probioticsmart.com	www.todounpl
www.bettafishstore.com	www.fashionista.com.br	www.kmart.com	www.raabe.de	www.totalinspr
www.biopoint.com.br	www.fastbacklink.de	www.konsonet.com	www.rachaelraystore.com	www.tradoria.c
www.biovitamineshop.de	www.foodnetworkstore.com	www.kosmetic.pl	www.ravensburg.de	www.tripwolf.c
www.bitmunk.com	www.franz.com	www.la-mousson.de	www.regiojatek.hu	www.tu-travels
www.bonvino.de	www.furniturefind.com	www.leetboss.com	www.renault.co.uk	www.tvstands.
www.bottleworld.de	www.gasparotto.biz	www.lis-og.com	www.robinsonoutdoors.net	www.universu
www.brandeer.ch	www.gingersjewelryonline.com	www.logicpass.ru	www.sachse-stollen.de	www.usclats-g
www.breastpumpdeals.com	www.glastradition-onlineshop.de	www.lojacarrion.com	www.sanderslaw.com	www.varejo-vlj
www.btrinfo.com.br	www.globalautoimports.com.br	www.lojacarrion.com.br	www.san-diego-dentistry.com	www.vedaboo
www.bunkersofa.com	www.gnowsis.com	www.lojamadameg.com.br	www.saveonvideo.com	www.voiiesans
www.buntegeschenke.de	www.golfhq.com	www.lojavirtual.rudel.com.br	www.schwitzen.com	www.volkswag
www.cafe-reisinger.at	www.goodboatshop.co.uk	www.lokool.com	www.sd-kyber.com	www.vx.valex.
www.capitalrefrigeracaodf.com.br	www.gorskisziak.pl	www.lovejoys-ltd.co.uk	www.sears.com	www.waffen-fi
www.cardgameshop.com	www.greatautodealersites.com	www.michaeliambertz.net	www.shop.friis-holm.dk	www.web.utan
www.cazzahair.com.br	www.gsmboutique.ro	www.mmmeeja.com	www.shopforia.com	www.webconti
www.cf.mixcontrol.pl	www.haar-shop.ch	www.modernchair.com	www.shop-matrixrom.ro	www.webno1.
www.cisema.ch	www.haengemattenshop.com	www.monsterclean.net	www.skybad.com	www.websitec
www.cloudofdata.com	www.hagemann24.de	www.mrwatch.com	www.skybad-de.de	www.wellenrei
www.colemanphotographix.com	www.heavy-liquid.com	www.msd-kilian.de	www.slindi.com	www.xtremeim
www.colibra.com	www.heels.com	www.netdental.com.br	www.smart-infosys.at	www.yoganaw
www.commercialloandirect.com	www.herbaman.com.ar	www.novadetox.co.uk	www.snookshop.com.br	www.zavita.de
www.connectors.de	www.hewettresearch.com	www.oetti.it	www.sponsorthisroundabout	
www.connex-filter.eu	www.hom.yoganaweb.com	www.olutools.com	www.sponsorthisroundabout.com	
www.cookware.com	www.hummel.si	www.onlinepharmaciescanada.com	www.sportenmagazin.net	
www.corsetsandcurves.com.au	www.hypotheekbelgieconsultant.be	www.ontosolutions.com	www.statsoft.com	

Fig. 5.11 List of Data sources included in GRDS

5.6.3   Data Providers Landscape

By observing the structured eCommerce data landscape (while building the GRDS),
we were able to categorize data publishers into three groups, based on their publishing
approach, usage pattern and data volume.

5.6.3.1   Large Size Retailers

This group includes large online eRetailers and retailers who are traditionally
premises-based and have only recently entered the eRetailing business. These
data sources provide detailed, rich offerings and product descriptions which are
useful for entity consolidation and interlinking with other datasets. Such companies include www.Volkswagen.com.uk, https://www.BestBuy.com, https://www.
Overstock.com, https://www.Oreilly.com, and www.Suitcase.com, to name a few.

5.6.3.2   Web Shops

A large number of semantic eCommerce adopters are small to medium Web shops
which offer their products and services mainly through Web channels. Most of these
Web shops use Web content management packages[11] such as Maganto[12], Oxid-
eSales,[13] WP 4 eCommerce,[14] osCommerce[15] and Joomla Virtuemart[16] to add RDFa
data in offer-related Web pages. This approach of embedding Semantic Web data in
existing Web pages works well for small and medium Web shops since no special
infrastructure arrangement is required in most cases because the semantic metadata
(data describing products and offers) is embedded within existing Web documents,
hence offering several benefits to both producers and consumers.

5.6.3.3   Data Service Providers (Data Spaces)

To leverage the benefits offered by semantic eCommerce data, businesses are offering
data services that are built on consolidated semantic repositories. Moreover, the
providers use APIs to access and transform proprietary data into RDF before making
them available through their repositories.

---

[11]A complete list of their references is available at http://www.ebusiness-unibw.org/wiki/
GoodRelationsShop_Software (retr., 15/10/2017).

[12]www.magentocommerce.com (retr., 15/11/2017).

[13]www.oxid-esales.com/ (retr., 23/11/2017).

[14]https://wordpress.org/extend/plugins/wp-e-commerce/ (retr., 21/10/2017).

[15]www.oscommerce.com/ (retr., 19/10/2017).

[16]https://virtuemart.net/ (retr., 07/11/2017).

5.6.4   Use of Different Namespace Analysis in GRDS

The availability of specific ontologies in the dataset and their usage intensity can be seen by querying the dataset and identifying the data sources using those ontologies. A different approach is adopted in reporting namespaces. Instead of counting the number of triples matching specified criteria, the percentage of the data sources that match the criteria available is reported. This approach provides less biased usage analysis, because it disregards the size of the implementer and looks at the number of data sources using it. For example, a large implementer such as https://www.BestBuy. com uses a term (e.g. `gr:contains`) to describe its two hundred thousand products and happens to be the only data source using this term in the dataset, hence this will count as only one instance of usage in the dataset. Table 5.2 lists the vocabularies present in the captured dataset along with the percentage of data sources using them.

In total, 48 namespaces are found in the dataset, of which 22 are listed in Table 5.2 and the others are excluded from the list. It is found that there are 12 in-house ontologies with no formal description, 4 with erroneous URIs and 7 namespaces representing W3C's formal specification such as RDF, RDFS, OWL, etc. The complete list of vocabularies found in the GRDS dataset is presented in [7]. The first four vocabularies in GRDS (see Table 5.2), next to *gr*, namely *vCard*, *foaf*, *Yahoo* and *dc* are, on average, used by 53% of the data sources to describe commonly used entities.

Table 5.2 List of vocabularies and their percentage in GRDS

Prefix	Namespace	% Data sources
Gr	http://purl.org/goodrelations/v1#	97.16
vCard	http://www.w3.org/2006/vcard/ns#	79.15
foaf	http://xmlns.com/foaf/0.1/	54.98
yahoo	http://search.yahoo.com/searchmonkey/commerce/	41.71
Dc	http://purl.org/dc/terms/	36.49
eCl@ss	http://www.ebusiness-unibw.org/ontologies/eclass/5.1.4/#	18.01
V	http://rdf.data-vocabulary.org	16.59
Og	http://opengraphprotocol.org/schema/	9.00
rev	http://purl.org/stuff/rev#	7.11
pto	http://www.productontology.org/id/	1.90
geo	http://www.w3.org/2003/01/geo/wgs84_pos#	0.95
Cc	http://creativecommons.org/ns#	0.95
frbr	http://vocab.org/frbr/core#	0.47
void	http://rdfs.org/ns/void#	0.47
sioc	http://rdfs.org/sioc/ns#	0.47
vso	http://purl.org/vso/ns#	0.47
coo	http://purl.org/coo/ns#	0.47
scovo	http://purl.org/NET/scovo#	0.47
comm	http://purl.org/commerce#	0.47
media	http://purl.org/media#	0.47

5.7 Empirical Analysis of Domain Ontology Usage

In this section, domain ontology usage is empirically analysed based on the GRDS dataset and the metrics defined in Sect. 5.4 as part of the EMP-AF. The computation of certain metrics defined for the empirical analysis required preprocessing to overcome the computational challenge. Before proceeding with the analysis, the preprocessing performed as part of the Empirical Analysis Module is discussed in the next section.

5.7.1 Preprocessing

In order to compute the metric values and gather the results of simple measures (computationally less expensive) such as concept instantiations, the presence of certain triple patterns and the use of different properties with a given pivotal concept are obtained by posing SPARQL queries to the dataset. However, for computationally complex operations such as traversal path, querying the dataset using the triple stores SPARQL endpoint does not offer a practical solution. Any query with more than three triple patterns in chain with fitter clauses fails to return the result set in a reasonable time. As a workaround, the dataset is exported into N-Triples (a line-delimited syntax for RDF graphs) format using Jena API [190] and nxparser API[17] is used to extract the paths fanning out from the pivotal entity. The list of paths is then used to compute the maximum and average path length. Additionally, the path steps of each path are generated and their frequency in the path list is updated to reflect the occurrences of each path step in the path's list.

To understand the use of label properties by the data publishers, two metrics are used, namely $Entity_{fl}$, $Entity_{dl}$ to measure the use of formal label properties and domain-ontology-specific label properties, respectively. Aside from RDFS, several ontologies have defined their own labelling properties which are often used together to provide the same contextual information but using different predicates. Publishers do this to provide support for vocabularies to make it easy for consumers, but it sometimes becomes an issue to decide which one to use while querying the data, from the consumers point of view. A few labelling properties which are formally defined as sub-properties (using `rdfs:subPropertyOf`) of `rdfs:label`, make it easy for the application to include all the labels available for an entity, if lightweight reasoning is supported. To make our analysis of labels more empirically grounded, the definition of $Entity_{fl}$ was relaxed to also include all the labelling properties which are sub-properties of `rdfs:label` and this includes: `foaf:name`, `skos:prefLabel`, `sioc:name` and `skos:prefLabel`. Another exception/ extension has been made to include dc:title, even though it is not defined as a sub-property of `rdfs:label`, but since it is one of the largely used [89] properties in LOD, it is included under $Entity_{fl}$. After relaxing the conditions, the following is the set of label properties as part of the formal labels:

---

[17]http://code.google.com/p/nxparser (retr., 12/11/2017).

$Formal Labels = \{$`foaf:name, skos:prefLabel, sioc:name,`
`dc:title`$\}$

To compute $Entity_{dl}$ for a given pivotal concept, a set of label attributes defined by the domain ontology is needed where the pivotal concept is the `rdfs:domain` of the label property. For the three pivotal concepts used in this analysis, the following is the set of domain labels. $Domain\,Labels_{gr:BusinessEntity} = \{$`gr:legalName`$\}$
$Domain Labels_{gr:Offering} = \{$`gr:condition, gr:category`$\}$
$Domain Labels_{gr:Product Or Service} = \{$`gr:category, gr:color,`
`gr:condition, gr:datatypeProductOrServiceProperty`$\}$

Based on the preprocessing approach discussed above, the SLG and the usage of each pivotal concept is analysed using the CUT metrics.

5.7.2 Analysing the Schema Link Graph (SLG)

Using the Schema Link Graph model, a graph representing all the ontologies available in the dataset was obtained, where the links reflect the co-usability of different ontologies. Figure 5.12 shows the links between the entities defined across various ontologies. The node size represents the degree of an ontology, which means the number of other ontologies linked with the ontology in further describing the entities available in the dataset. For example, the *foaf* node has a degree value of 7 which means that the *foaf* resources are further linked with *dc, frbr, vso, vCard, pto, gr* and *v* resources. In the Schema Link Graph, the average node degree is 4.12 with a standard deviation 3.61, which shows that the degree distribution ostensibly follows the Power Law distribution [61]. However, the average degree distribution in the Schema Link Graph is encouraging because it reflects a good co-usability factor in the dataset. After analysing the use of different vocabularies and the linking of entities over vocabularies, the domain ontology usage is examined in the next section in a more detailed fashion, to understand the data and knowledge patterns available in the dataset.

5.7.3 Analysing the Concept Usage Template (CUT) and Labelling

In order to carry out the empirical analysis of the domain ontology, it is important to identify the pivotal concepts which represent the core entity in the domain conceptualized by the domain ontology. While there are some advanced approaches [293] available which can be employed to automatically find the key concepts of the domain ontology, the `gr:BusinessEntity`, `gr:Offering` and `gr ProductOrService` pivotal concepts, introduced in Sect. 5.5.2 were used.

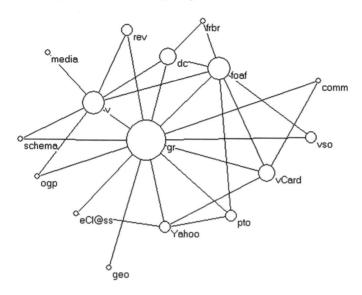

Fig. 5.12 Schema Link Graph (SLG) in GRDS

Table 5.3 CUT of gr:BusinessEntity

Entity	gr:BusinessEntity
Instantiation	54,542
Vocabs	vCard, gr, foaf, yahoo, v and schema
Object properties	vCard:adr, vCard:email, vCard:url, yahoo:image, gr:offers, gr:hasPOS, foaf:logo, foaf:homepage, foaf:maker,, foaf:page, gr:hasOpeningHourSpecification, foaf:depiction
Attributes usage	vCard:fn, vCard:tel, vCard:email, vCard:organization-name, vCard:fax, vCard:adr, vCard:Tel, gr:hasISICv4, gr:legalName, v:name, v:pricerange, v:category, foaf:maker, yahoo:seatingOptions, yahoo:cuisine, yahoo:features, yahoo:smoking, yahoo:serviceOptions, yahoo:mealOptions, yahoo:priceRange, yahoo:hoursOfOperation, schema:postalCode, schema:addressLocality, schema:streetAddress, schema:telephone
Class usage	vCard:VCard, cVard:org, yahoo:Business, yahoo:Restaurant, gr:BusinessEntityType, comm.:Business, v:Organization
Interlinking	rdfs:seeAlso, owl:sameAs

5.7.3.1 gr:BusinessEntity Analysis

In GRO, gr:BusinessEntity represents a business organization (or any individual) which intends to offer or seek products on the web. The RDF usage based on the CUT is examined, then the available paths and labels provided with the entities of this concept are discussed. Table 5.3 provides the analysis results for the gr:BusinessEntity concept. In our dataset, $CI(gr : BusinessEntity)$ i.e.

Eq. 5.1 is 789,440 entities in total and of these, 54,542 are of the type `gr:Business Entity` concept. This means that 6.9% of the entities are of this type in the GRDS. From the *Vocab* (Eq. 5.3) set, the co-usage of different vocabularies in the entity description can be seen. The list of object properties provides an approximation of the relationships the entity has and provides substantial evidence about the discoverable related entities in the knowledge base. By looking at the object properties, it can easily be seen that this pivotal business entity is described with its location address and contact-related details. In addition to this relationship, attribute usage provides all the attributes used to generate textual information about the entity. In the RDF data, it is presumed that all the resources are identified with URIs which, when dereferenced, will return human readable information about the resource. Interestingly, in attribute usage ($Attri(C)$) found the use of several attributes which are from schema.org[18] and are not valid URIs. This also indicates the adoption and use of non-semantic schema in RDF data which is believed to be a good sign as far as the burgeoning of structured data on the Web is concerned, although the semantic aspect is being ignored.[19]

Class Usage ($ClassUsage(C)$), which lists the other classes of which the entity is a member, returns seven other classes. This tells us that one or more entities of the `gr:BusinessEntity` class in this dataset also has membership of seven other classes. This membership relationship information provides the intrinsic overlapping in the conceptualization of different concepts which have several aspects in common, but not essentially the same interpretation in cross domains. To promote information interoperability on the Web, the identification of related but different concepts in the knowledge base facilitates alignment between different concepts in the ontology mapping process. We believe that related concepts often maintain an elusive relationship, requiring more diverse mapping predicates to capture the natural linkages between disparate concepts instead of using a mapping predicate with strong semantics, i.e. `owl:equivalentClass` [19].

In interlinking, the information related to the linking of similar but disparate entities is captured. This includes the link base and the equivalence links indicating that different URIs are, in fact, referring to the same resource or entity. We found the use of two interlinking properties for the entities in the dataset, namely `owl:sameAs` and `rdfs:seeAlso`. The `rdfs:seeAlso` provides very little information about the resource it links to but is a standard Semantic Web method of linking hypertext to provide reference to additional resources or documents. The last component of CUT measures the use of label properties in the dataset. As mentioned earlier, $Entity_{fl}$ and $Entity_{dl}$ metrics are used to measure the use of formal label predicates and the domain ontology-specific label properties, respectively. Focusing on this pivotal concept, 32% of entities have used the label properties with the following values for these two metrics:

---

[18]http://schema.org (retr., 02/11/2017).

[19]On a side note, there has been a community effort to map schema.org terms, with their semantic version published at http://schema.rdfs.org/mappings.html.

$Entity_{fl}$ = 1,703 (9% of entities have used formal labels)
$Entity_{dl}$ = 17,146 (91% of entities have used domain labels)

One of the most obvious and surprising findings is the dominance of the domain label predicates over the formal labels. Contrary to the previous findings in [89, 186] and the general presumption that formal labels are more frequently used, the dominance of domain ontology-specific label properties can be seen in our experiment. This also signifies that information (data) publishers prefer to provide specialized label properties to help consumers access less ambiguous contextual information, which is useful for querying and interface presentation.

5.7.3.2 gr:Offering Analysis

gr:Offering is the concept which enables business entities to publish their offers on the Web, either for selling or buying products. Table 5.4 presents the CUT for the gr:Offering pivotal concept. In RDF usage, an interesting finding is the use of different but related vocabularies to semantically describe offering-related information. Three vocabularies which supplement offering information, namely *media*, *rev* and *comm* are included; however, two names which are included in the gr:BusinessEntity concept, *vCard* and schema vocabularies have been excluded. Similar to the previous concept, the use of different predicates from different vocabularies used to provide the offering description can be seen in both Object Property (Eq. 5.4) and Attribute Usage (Eq. 5.5). Another interesting finding is the

Table 5.4 CUT of gr:Offering

Entity	gr:Offering
Instantiation	61,330
Vocabs	gr, foaf, v, comm, media, rev, yahoo
Object properties	gr:availableAtOrFrom, gr:hasBusinessFunction, gr:eligibleCustomerTypes, gr:acceptedPaymentMethods, gr:availableDeliveryMethods, gr:includesObject, gr:hasPriceSpecification, gr:hasWarrantyPromise, gr:includes, gr:hasManufacturer, gr:hasInventoryLevel, gr:hasBrand, foaf:page, foaf:depiction, foaf:thumbnail, yahoo:media/image, yahoo:product/specification, yahoo:product/manufacturer, v:url, v:photo, v:hasReview, media:depiction, media:sample, media:contains, rev:hasReview
Attributes usage	gr:validFrom, gr:validThrough, gr:eligibleRegions, gr:hasStockKeepingUnit, gr:availabilityStarts, gr:hasEAN_UCC-13, gr:description, gr:name, gr:condition, gr:hasMPN, gr:BusinessEntity, gr:hasCurrency, rdfs:title, rdfs:comments, dc:description, dc:title, dc:contributor, dc:date, dc:type, dc:duration, dc:position, v:name, v:description, v:price, v:category, v:brand, ogp:image, ogp:type, ogp:site_name, ogp:title, ogp:url
Class usage	v:Product, media: Album, media:Recording note: I have found around 26 product types defined by http://www.productontology.org/
Interlinking	rdfs:seeAlso, owl:sameAs

use of product vocabularies to describe the products being offered; therefore, the use of different concepts defined in product ontology as part of the Class Usage can also be seen. Since the list is long, this chapter only provides the concepts used from the pro-vocabulary. The use of interlinking predicates is the same as the previous pivotal concept, and one can readily assume that these two predicates are consistent across all key concepts and entities.

Next, the use of label properties by the entities of the gr:Offering type are analysed. Of 61330 entities, 11% used labelling properties with the following distribution:

$$Entity_{fl} = 4,171 \ (62\% \text{ of entities used formal labels}) \ Entity_{dl} = 2,610 \ (38\% \text{ of entities used domain labels}).$$

5.7.3.3   gr:ProductOrService Analysis

In GRO, a lightweight description of the products being offered is presented through gr:ProductOrService and three of its sub-classes.

Table 5.5 shows the usage summary for the gr:ProductOrService concept. In total, there are roughly 38,000 entities defined as 'type of product'. Since in GRO, product-related concepts are arranged in a taxonomical hierarchy to allow users to specify the exact nature of the product being offered, the subsumption axiom is used to include all the instances belonging to the super concept. Vocabulary usage for product and offering is almost identical and the entities of both concepts use the same vocabularies to describe the instances. One important improvement to Class Usage, compared with our previous study [8] is that most new eCommerce data pub-

Table 5.5 CUT of gr:ProductOrService

Entity	gr:ProductOrService
Instantiation	37,996
Vocabs	gr, foaf, yahoo, v, vso, eCl@ss, pto
Object properties	gr:hasMakeAndModel, gr:hasInventoryLevel, gr:hasManufacturer, gr:description, gr:depth, gr:height, gr:weight, gr:width, vso:mileageFromOdometer, gr:hasBusinessFunction, gr:hasMakeOrModel, gr:hasBrand, gr:hasPriceSpecification, foaf:depiction, foaf:thumbnail, foaf:page, foaf:logo, rev:hasReview, v:hasReview, vso:bodyStyle, vso:engineDisplacement, vso:gearsTotal, vso:previousOwners, gr:name, vso:transmission, vso:fuelType, vso:feature (*note: there are several in-house developed ontologies to describe product attributes*)
Attributes usage	gr:description, gr:hasStockKeepingUnit, gr:hasEAN_UCC-13, gr:name, gr:hasMPN, gr:condition, gr:category, vso:modelDate, vso:VIN, vso:color, vso:engineName, vso:rentalUsage
Class usage	eCl@ss, v:Product, yahoo:Product, vso:Automobile (*note:* http://www.productontology.org *has hundreds of classes which are used in dataset for describing high level product type/category*)
Interlinking	rdfs:seeAlso

lishers now use product ontologies to describe their products. For example, in our dataset, more than 100 concepts of *pto* are used to specify the types of product being offered. In interlinking, the usage of rdfs:seeAlso predicate is seen, however, there is no usage instance of the `owl:sameAs` predicate. Possible reasons for the temporary nonexistence of this predicate in product instances is firstly that although product ontologies have recently begun to emerge, these ontologies do not offer rich product descriptions such as covering the qualitative and quantitative properties of products, and secondly, that `owl:sameAs` interlinking is algorithmically complex and less effective, and preferably done through social engagement.[20] Pertaining to the use of label properties with product instances, the label metric values are as follows:

$Entity_{fl} = 30,379$ (99.05% of entities are using formal labels) $Entity_{dl} = 360$ (0.95% of entities are using domain labels)

In the product pivotal concept, 30,739 entities have labels attached to the instances, which means that 80% of the entities offer textual descriptions to provide human readable descriptions of the product. Of these 80%, only 0.95% of the entities provide domain label properties and 99.05% provide formal labels, which is quite a different trend compared to the above two pivotal concepts. As mentioned earlier, GRO provides only high level concepts to identify the product but recommends using product ontologies such as *eCl@ss* and *pto* to provide semantic descriptions of products, therefore, there is little or negligible use of domain ontology-specific labels.

5.7.4 Analysing Knowledge Patterns (Traversal Path)

Referring to Sect. 5.4.5, traversal path metrics are defined to understand the available knowledge patterns in the dataset by constructing traversal paths and computing the strength of the path steps in those paths. The number of traversal paths in the dataset, originating from each pivotal concept, is presented in Table 5.6.

Table 5.6 shows the number of unique paths which exist for each pivotal concept. To recap, in traversal paths, all the unique paths originating (fanning out) from the given pivotal concept are calculated. This provides the data and schema level patterns available in the knowledge base. Since gr:BusinessEntity is considered a kind of root (not in the literal sense) concept, it can be seen that it has the largest maximum traversal path length. Similarly, gr:ProductOrService, being the later concept in the ontological model, has the lowest maximum length. Interestingly, there is little significant deviation in the average path length, which indicates that even though gr:BusinessEntity has the maximum path length on average, all the pivotal concepts

---

[20]In a keynote speech at ISWC2011 (http://www.cs.vu.nl/~frankh/spool/ISWC2011Keynote/; retr., 28/11/2017), Frank van Harmelen mentioned the role of social engagement being more effective than an algorithmic approach in interlinking entities.

Table 5.6 Traversal path of all three pivotal concepts

	gr:BusinessEntity	gr:Offering	gr:ProductOrService
Number of unique paths	12,245	14,871	2,453
Maximum path length	6	4	3
Average path length	3.12	2.78	2.13

Table 5.7 Path Steps frequency in Traversal Path

Path step	Frequency
gr:Offering gr:hasBusinessFunction gr:BusinessFunction	51928
gr:Offering gr:hasPriceSpecification gr:PriceSpecification	34659
gr:Offering gr:includesObject gr:TypeAndQuantityNode	29038
gr:Offering gr:availableAtOrFrom gr:Location	24914
gr:Offering gr:hasManufacturer gr:BusinessEntity	19430
gr:Offering gr:eligibleCustomerTypes gr:BusinessEntityType	15906
gr:SomeItems gr:hasMakeAndModel gr:ProductOrServiceModel	7168
gr:Offering gr:availableDeliveryMethods gr:DeliveryMethod	5462
gr:Offering gr:hasWarrantyPromise gr:WarrantyPromise	4090
gr:BusinessEntity gr:offers gr:Offering	2398
gr:BusinessEntity vCard:adr vCard:Address	2385
gr:OpeningHoursSpecification gr:hasOpeningHoursDayOfWeek gr:DayOfWeek	1953
gr:Offering gr:includes gr:ProductOrService	1814
gr:Location gr:hasOpeningHoursSpecification gr:OpeningHoursSpecification	1025
gr:BusinessEntity gr:hasPOS gr:Location	598
gr:Offering media:contains v:Product	514
gr:BusinessFuntion gr:hasBrand gr:Brand	265
gr:Offering media:contains media:Recording	218
gr:BusinessEntity vCard:url owl:Ontology	182
gr:WarrantyPromise gr:hasWarrantyScope gr:WarrantyScope	19
gr:DayOfWeek gr:hasNext gr:DayOfWeek	7
gr:DayOfWeek gr:hasPrevious gr:DayOfWeek	7
gr:Offering rev:hasReview rev:Review	4

have a close average path length. This kind of insight into data and schema patterns
and the depth in triple chaining patterns helps in planning data management including storage, querying and reasoning. To understand the triple patterns available in
traversal paths, the following table lists the dominant path steps extracted from the
paths with their frequency.

Table 5.7 lists the dominant path steps with the frequency found in traversal paths.
This provides a snapshot of the terminological knowledge and the schema level triples
available in the dataset. This and the traversal path information, which provides the

summary of the knowledge base, helps in generating the SPARQL query template to access domain-related knowledge from any dataset. However, note that while this provides a complete set of terminologies used in the dataset, not necessarily all entities use these terms, therefore certain terms in the automatic query generation process need to be optional. To support more effective automatic query generation, based on the summary above, one can consider attaching frequency to each term to provide an estimate of distribution. In the next section, the empirical analysis obtained using the EMP-AF is evaluated using some of the requirements discussed in the introduction.

5.8 Empirical Analysis Evaluation

There are different types of users, each of whom may have their own requirements pertaining to the required understanding on the use of ontologies, as mentioned in Sect. 1.7. The aim of empirical analysis is to obtain a detailed insight into the use of domain ontologies on the Web. In the aspect analysis phase, key aspects which can provide broader visibility of the adoption, uptake and usage of domain ontologies are considered to define the metrics for investigation. The following subsection will analyse how these results help to address a few of the questions raised in the introductory section, using the results obtained by employing the developed metrics.

5.8.1 Scenario 1: Application Developers Need to Know How a Given Ontology Is Being Used

For Semantic Web application developers, it is important to know the nature, structure and volume of data available to them for the application. By using the EMP-AF, there are several sub-requirements which can be identified to provide precise information to developers. These precise requirements are described in the following sub-cases.

5.8.1.1 Case 1: What Terminological Knowledge Is Available for Application Consumption?

Terminological knowledge, which refers to the use of terms (vocabularies) defined by ontologies, is important because it provides a representation and description of the entities involved in the given domain. Application developers using this information can prepare generic queries to access the data or prepare the interface based on the available (ontological) conceptual elements. The Concept Usage Template (CUT), which captures all the terminological knowledge attached to the concept, provides a unified source of information to the developer (as well as to other types of ontology users) for preparing the data access layer. For example, Table 5.3 shows how gr:BusinessEntity concept is generally being used and provides specific details on how many instances of this concept are present (i.e. 54,542), what other entities it is connected to, and what relationships it uses. As shown in Table 5.3, *vCard:adr, vCard:email, vCard:url, yahoo:image, gr:offers, gr:hasPOS, foaf:logo, foaf:homepage, foaf:maker,, foaf:page, gr:hasOpeningHourSpecification, and foaf:depiction* relationships (object properties) are used to provide relevant details for the instances of the concept.

5.8.1.2 Case 2: What Common Data and Knowledge Patterns Are Available?

From a data management and processing point of view, it is important to know the different types of patterns being followed in the dataset (or usage in general). Information regarding the patterns not only helps in generating prototypical queries but also assists in strategizing the index for efficient information retrieval and storage. Traversal paths and their frequency identify the presence of knowledge patterns and their frequency in the dataset. For example, in Table 5.7, it can be seen that the knowledge pattern which dominates the whole dataset (indicating that the majority of data publishers have published this piece of knowledge) is *(gr:Offering –> gr:hasBusinessFunction –> gr:BusinessFunction)* and this pattern has 51,928 occurrences in the dataset, whereas at the other extreme, *(gr:Offering –> rev:hasReview –> rev:Review)* patterns have the least number of occurrences four.

5.8.1.3 Case 3: How Are Entities Being Annotated or Textually Described?

Information regarding the use of different properties to provide textual description to entities is very helpful for developers (as well as to other users) in a number of ways. For example, knowing which textual or annotative property is being used helps developers to design a user interface in which the "human readable" description of the entities is displayed, rather than showing the URI which is opaque in describing what an entity is, and is not reader friendly. Additionally, the information regarding which labelling properties are being used, whether of standard vocabularies (such as RDF, RDFS, OWL) or of other vocabularies, including domain ontologies (such as DC, FOAF, GR), helps to develop an interface that displays information that is machine accessible but also human readable. In the case of the *gr: BusinessEntity* concept, almost 91% of data publishers have used the ***domain labels*** to provide a textual description of the entity, and the labelling property used for this concept is `gr:legalName` which provides a human readable name of the business entity.

5.8.2 Scenario 2: Data Publishers Need to Know What Is Being Used to Semantically Describe Domain-Specific Information

As mentioned in Sect. 5.1, it is recommended that data publishers, wherever possible, reuse ontologies instead of developing new terms or ontologies, the reason being that the more an ontology is reused, the more value it has in terms of perceived utility. For data publishers, therefore, it is desirable to know how a given entity is being described and what ontologies are being used. By using EMP-AF, two such requirements are analysed and presented to data publishers to provide them with the required insight.

5.8.2.1 Case 1: How Is a Company (or Business) Being Described and What Attributes Are Being Used?

It is very important for any business to provide a semantic description of their business to make their products or services discoverable by agents/clients. The best approach is to understand how such information is published by others and what the prevailing structure on the Web is. The dominant structure provides the template which can then be used for publishing Semantic Web data on the Web. EMP-AF provides CUT to capture this structure and assists the data publisher in their publishing process. Table 5.3 provides the prevalent semantic description of *gr:BusinessEntity* (which conceptualizes the concept of a company/business) and can be used by data publishers to describe their company.

Attribute usage (the fifth row of Table 5.3) provides a list of datatype properties being used by others, helping data publishers to know what attributes and which terms are being used to describe a company. Specific to the case study considered in this chapter, a few of the attributes (for a complete list see Table 5.3) used are: *gr:legalName, vCard:fax, vCard:adr, vCard:Tel, schema:postalCode, schema: addressLocality,* and *schema:streetAddress*

5.8.2.2 Case 2: What Other Entities Are a Company (Entity) Linked To?

For data publishers, it is important to know how a given entity is being linked with other entities and what relationships are being used. The availability of such information helps data publishers specifically and others generally to know in what dimensions an entity is being described and interlinked with other ontologies. For example, is the company only being described to provide address-related information or is the company's product-and-service-related information also being described? The CUT metric of EMP-AF provides sub-metrics to obtain the specific details of concept usage. One of the sub-metrics is $ObjectPro$ which captures the relationships the pivotal concept has with other resources (entities). Table 5.3 (fourth row) provides a list of relationships (object properties) the business/company type entity has with other entities. A few of the relationships being used (see Table 5.3 for complete list) are: *gr:offers, gr:hasPOS, foaf:homepage, foaf:maker, foaf:page, gr:hasOpeningHourSpecification*. It can be seen that other data publishers have provided information pertaining to the branches a company has, offers relating to its products/services, and the address of the company's homepage.

5.9 Recapitulation

In this chapter, the EMP-AF was presented to perform empirical analysis on the use of domain ontologies on the Web. The developed metrics were used on the dataset to analyse how the domain ontology (GoodRelations, in this case) is being used and how its key concepts are described. The insights obtained assist in addressing the needs and requirements of different types of ontology users in order to make effective and efficient use of the available Semantic Web data.

In the next chapter, which also implements the investigation phase of the OUSAF, the use of domain ontologies are quantitatively analysed. The quantitative analysis provides the quantitative measures to help in further realizing the benefits of Ontology Usage Analysis.

Chapter 6
Investigation Phase: Quantitative Analysis of Domain Ontology Usage (QUA-AF)

Focus of the Chapter

6.1 Introduction

[1]In the previous chapter, the EMP-AF was proposed to perform an empirical analysis of domain ontology usage. The empirical analysis, through its observed factors such as the relationship between different ontologies based on an entity's semantic description, ontology component usage, contextual description, and provision and availability of knowledge patterns, helps us to understand the uptake and adoption of domain ontologies on the Web. In other words, it gives a comprehensive analysis of the "usage" aspect of a domain ontology and its components. While the insights obtained through EMP-AF highlight the key aspects of the *usage* dimension, two other dimensions warrant consideration to fully realize the perceived benefits of

---

[1]Parts of this chapter have been republished with permission from:

1. Jamshaid Ashraf, Omar Khadeer Hussain Farookh Khadeer Hussain, A Framework for Measuring Ontology Usage on the Web, The Computer Journal, 2013, Volume 56, Issue 9, pp. 1083–1101, by permission of OUP.

2. John Wiley & Sons from Empirical analysis of domain ontology usage on the Web: eCommerce domain in focus, Jamshaid Ashraf, Omar Khadeer Hussain, Farookh Khadeer Hussain, Volume 26, Issue 5, Copyright 2013 John Wiley & Sons, Ltd; permission conveyed through Copyright clearance centre.

3. John Wiley & Sons from Making sense from Big RDF Data: OUSAF for measuring ontology usage, Jamshaid Ashraf, Omar Khadeer Hussain, Farookh Khadeer Hussain, Volume 45, Issue 8, Copyright 2014 John Wiley & Sons, Ltd; permission conveyed through Copyright clearance centre.

© Springer International Publishing AG 2018

J. Ashraf et al., *Measuring and Analysing the Use of Ontologies*, Studies in Computational Intelligence 767, https://doi.org/10.1007/978-3-319-75681-3_6

Ontology Usage Analysis (OUA), as mentioned in Chap. 3, and to enable the quantitative analysis of OUA. These are the "technology" and "business" dimensions, which also have a direct relationship with ontology adoption and usage.

The **technology dimension** captures the technical antology and its components, such as the richness of the structural representation that assists in the usage of its components by different users. It symbolizes the conceptual model, which includes the structural characteristics of ontologies, and the formal model, which includes the formalization of the conceptualized model. In other words, it considers the design, structural and functional aspects of ontologies to capture those characteristics in the OUA.

The **business dimension** embodies the impetus or commercial advantage (be it monetary or technological) received directly or indirectly by end users through the use of ontologies. In other words, it quantifies the incentives available to the ontology user or the ontology itself because of its recognition, popularity and dominance. It is important to consider these two dimensions, along with the **usage dimension** that provides an insight into the use of domain ontologies in real world settings, to have a comprehensive multi-dimensional insight to ontology usage and its adoption in the real world. Considering these three dimensions together also closely aligns with the "usage model" presented in [234], in which the author states that any compelling product is found at the intersection of "business", "usage", and "technology" dimensions, as shown in Fig. 6.1. In the context of OUA, ontologies being the engineering artefact are considered as "product" and their usefulness is measured through the three dimensions of "business" being the actual (quantified) value received through the use of ontologies, "usage" being the use of the product in the real world, and "technology" being the formal model behind the development of ontologies. To analyse domain ontology usage quantitatively, ontologies need to be analysed from the dimensions of **technology**, **usage**, and **business**. Each dimension covers a different aspect of ontology usage analysis, as described below:

1. **Measure the characteristics of an ontology and its components that assist in its usage (technology dimension)**: To comprehensively understand how ontologies are being used, and what exactly is being used, it is important to understand

Fig. 6.1 Usage model [234]

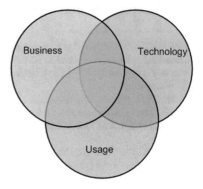

the characteristics of the conceptual model, its structure and components. In particular, it is important to measure how different concepts and relationships are defined in an ontology and their semantic description in the ontological model. In other words, the technology dimension measures the richness of ontology components which provides structural insight into how a given ontology is modelled and how the semantics are represented.

2. **Measure the use of an ontology and its components (usage dimension)**: This dimension measures the use of ontological components such as concepts, relationships and attributes. The measure helps in understanding how ontologies are used in real world settings.

3. **Measure the driving factors behind ontology adoption (business dimension)**: To gain a comprehensive insight into the use of ontologies and their components, it is important to identify and incorporate the driving factors behind the adoption of the ontologies. This dimension measures the benefits that are realized by users as a result of using an ontology.

To quantify these measures, a mechanism is required to compute and evaluate each dimension in order to undertake a comprehensive analysis of ontology usage. Therefore, quantitatively analyse the use of domain ontologies considering the above-mentioned requirements and dimensions, the **QUAntitative Analysis Framework (QUA-AF)** is proposed in this chapter. The rest of the chapter is organized as follows. Section 6.2 presents the QUA-AF and its three phases: the data collection phase, the computation phase and the application phase. It also describes the sequence of the set of activities carried out in the QUA-AF. Section 6.3 presents the metrics defined for each dimension to quantitatively analyse domain ontology usage. In Sect. 6.4, a case study focusing on the domain of eCommerce is presented which will be used in the rest of the chapter to explain the working of the QUA-AF and the interpretations of the results obtained from it. In Sects. 6.5 and 6.6, GoodRelations and FOAF ontologies (from the case study presented in Sect. 6.4) are quantitatively analysed using the QUA-AF. The evaluation of the framework on the analysed domain ontologies is discussed in Sects. 6.7 and 6.8 concludes the chapter.

6.2 QUAntitative Analysis Framework (QUA-AF)

The proposed **QUAntitative Analysis Framework (QUA-AF)** comprises three phases: the *data collection phase*, the *computation phase*, and the *application phase*, as shown in Fig. 6.2. The data collection phase is responsible for collecting the desired data in the required format from the various sources in order to perform the analysis of each dimension. In the computation phase, different sets of metrics are defined to analyse the ontology usage in each dimension. In the application phase, the results obtained are converted into actionable information. In the next subsection, the objective, technical aspects and working of each phase are presented in detail.

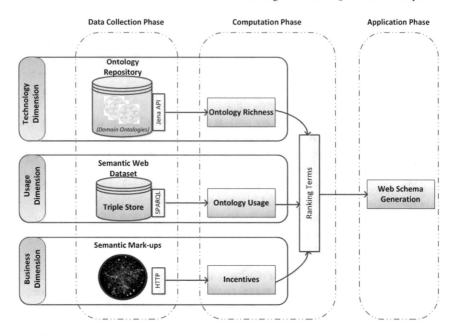

Fig. 6.2 Quantitative analysis framework for ontology usage analysis

6.2.1 Data Collection Phase

As mentioned earlier, the data required by the QUA-AF is collected in the data collection phase. Each dimension to be considered in quantitative analysis requires a different type of information to measure the aspects involved in it. As shown in Fig. 6.2, this is achieved by having a number of repositories to provide the dimension-specific data for computation purposes as follows:

6.2.1.1 Ontology Repository

The ontology repository collects the data to perform the analysis related to the *technology dimension*. Since the technology dimension captures and quantifies the design and structural characteristics of ontologies which assist in its adoption, the ontology repository hosts (stores) the authoritative representation of domain ontologies. The authoritative representation of an ontology includes ontology documentation, ontology formal conceptualization and metadata about the ontology. Generally, the main sources of this information are ontology libraries [83] which maintain the databases of different ontologies.

The use of existing ontology libraries for the QUA-AF raises two issues, however. First, libraries (like OntoServer [277], ONION [106], and Cupboard [67]), even though they are complete systems, are computationally expensive considering the need at hand because they require various pre-processing operations such as boot-strapping, meta-data entry, etc. Considering them therefore becomes an elaborate and configuration-extensive choice, leading to an increase in complexity. Second, most online ontology libraries are application-specific (such as OBO Foundry [238], which contains biological and biomedical domain-specific ontologies) and are there-fore limited in offering ontologies from diverse domains, which makes them less applicable for our case. To avoid such drawbacks, a local repository of different domain ontologies is maintained for the QUA-AF.

6.2.1.2 Semantic Web (RDF) Data

The Semantic Web Dataset collects the data for analysis related to the *usage dimen-sion*. In order to measure the use of an ontology and its components in a real world setting, RDF data is crawled from the Web, comprising published structured data described using semantic markups. The required dataset, which comprises real instance data annotated using domain ontologies, is crawled and maintained in triple stores to obtain the Semantic Web (RDF) data published on the Web.

6.2.1.3 Semantic Markup Repository

The Semantic Markup repository collects the data for analysis related to the *business dimension*. In order to identify the impetus which encourages data publishers (users) to publish semantically annotated structured data on the Web, a repository is needed to maintain the list of semantic markups supported by the various search engines that assist them in the identification and classification of information. The Semantic Markup repository needs to list all the terms being used which are recognized or supported by search engines, either while crawling the data or when being used as canonical terms to describe entities. This data is then used to measure the incentives of different vocabularies.

6.2.2 Computation Phase

The computation phase (See Fig. 6.2) of the QUA-AF focuses on performing quanti-tative analysis of domain ontology usage by computing different measurements for each dimension. This phase comprises three modules, each focusing on one of the dimensions described in the following subsections.

6.2.2.1   Ontology Richness Module

The ontology richness module determines the analysis related to the technology dimension of ontologies. In this module, the richness of ontology components such as concepts and relationships are measured and quantified to represent the technology dimension. This module accesses the ontology's authoritative documentation stored in the ontology repository to measure its typological and structural characteristics. For the computation of this information (conceptual model richness), Jena API [52] is used to access the ontologies and construct the graph model to measure different properties. Metrics are defined to measure the **concept richness**, **relationship richness**, and **attribute richness**. The metrics defined for the ontology richness module are described in detail in Sect. 6.3.1.

6.2.2.2   Ontology Usage Module

The ontology usage module determines the analysis related to the usage dimension of ontologies. It measures how a domain ontology and its components are being used in a real world setting. In measuring the use of different ontology components, it needs to consider the axioms available in the ontology to understand the implied usage of the terms defined in the ontologies. Semantic Web data comprising real world data published on the Web, and annotated using domain ontologies, is used for the computation of usage. Using Semantic Web data, this module defines metrics to measure **concept usage**, **relationship usage**, and **attribute usage**. The metrics defined for the ontology usage module are described in detail in Sect. 6.3.2.

6.2.2.3   Incentive Module

The incentive module determines the analysis related to the business dimension. This module captures the commercial advantages available to Semantic Web data publishers. It attempts to recognize the use of different semantic mark-ups (concepts) by the data-consuming applications (search engine, for example) and matching them with the terminological knowledge of different ontologies. The matching of semantic mark-ups with terms defined by ontologies is being considered a convincing motivational factor behind their adoption. It evaluates the available support for various ontologies by search engines (or other applications such as RDF triple store, semantic reasoner) and gives weight to those terms accordingly. To evaluate the support available in different search engines, manual effort is required to prepare the list of terms being supported by the engine.

As mentioned earlier, the business dimension refers to the commercial incentives or advantages being received by users through the use of ontologies. However, as mentioned in Chap. 1, it is difficult to quantify the commercial benefit due to the lack of studies or statistics in this regard, since we are still in the early stages of Semantic Web technology usage and adoption. Nevertheless, in this book we consider it a key factor in fostering the growth and adoption of vocabularies and view it as one of the "driving factors" for early adoption. Two of the other driving factors are the incentives available to structured data publishers as a result of publishing their data using ontologies and the support available for an ontology/vocabulary in Semantic Web applications and tools.

Using the Semantic markup list, this module defines the **incentive** metric to measure the available commercial incentive for domain ontologies.

The metric defined for the Incentive module is formally described in Sect. 6.3.3.

6.2.2.4 Ranking Different Measures

Once the analysis of each module has been completed, the results are combined to obtain a consolidated quantitative value that represents ontology usage. Each dimension in QUA-AF contains different metrics and involves different aspects of the ontology, therefore to obtain a unified observation of usage, the analysis output of each dimension is weighted according to its preference to generate a consolidated value. The final usage values are then ranked to obtain an ordering list, based on the users requirements.

The ranking approach used in the QUA-AF is formally described in Sect. 6.3.4

6.2.3 Application Phase

The application phase of the QUA-AF implements a use case to represent the obtained result. The use case scenario highlights the need for a consolidated Web Schema representing the information for a particular application area. The Web Schema that is generated is based on ontology usage analysis to capture the prevalent and prominent data usage patterns which can be then used by other data publishers. Therefore, based on the identified requirements of the use case scenario, the QUA-AF constructs the Web Schema and captures the terminological knowledge representing the information specific to a given application area.

In the next subsection, the set of sequential activities carried out by the QUA-AF is presented.

6.2.4 Sequence of QUA-AF Activities

As mentioned in Sect. 6.2, the QUA-AF comprises three phases: the data collection phase, the computation phase, and the application phase. Each phase involves a certain number of activities to carry out the required functionality and operation. The set of activities and their sequence followed in the QUA-AF is depicted in Fig. 6.3 and is described below.

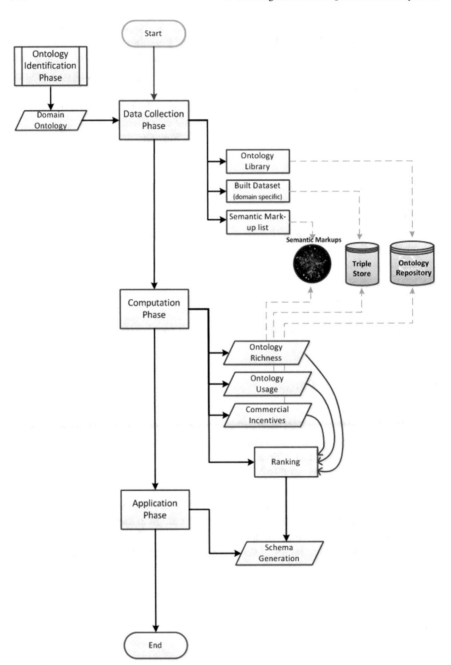

Fig. 6.3 Flow of activities in QUA-AF

- In the data collection phase, for each dimension, the following activities are performed.

  - To measure the technology dimension of ontologies, an <u>ontology repository is built</u> to store the domain ontologies along with their authoritative documentation.
  - To measure the usage dimension of ontologies, a <u>Semantic Web dataset is built</u> to store the RDF data published on the Web. The dataset is refreshed with new crawled data.
  - To measure the business dimension of ontologies, a list of <u>Semantic Markups supported by different search engines</u> is maintained.

- In the computation phase, the following activities are performed to measure the aspects of each dimension.

  - To measure ontology usage from the technical dimension, the <u>ontology richness module is defined</u>. The module contains the following metrics:
    <u>Concept richness</u> to measure the structural and typological characteristics of concepts.
    <u>Relationship richness</u> to measure the structural and typological characteristics of object properties (relationships).
    <u>Attribute richness</u> to measure the structural characteristic of datatype properties (attributes).
  - To measure ontology usage from the usage dimension, the <u>Ontology Usage module is defined</u>. The module contains the following metrics:
    <u>Concept usage</u> metric to measure the use of the concept.
    <u>Relationship usage</u> metric to measure the use of relationships.
    <u>Attribute usage</u> metric to measure the use of datatype properties (attribute).
  - To measure ontology usage from the business dimension, the <u>incentive module is defined</u>. The incentive module defines the incentive metric to measure the commercial incentives available to the user as a result of using the ontology.
  - Measures obtained in each module <u>are consolidated using a weight factor</u> to rank the ontologies and their components.

- In the application phase, the obtained quantified analysis is used to <u>construct the Web Schema</u>.

6.3 Metrics for Quantifying Dimensions for OUA in QUA-AF

In this section, the metrics to measure each dimension required for the quantitative analysis of domain ontology usage are defined. The metrics defined to measure the ontology from the various dimensions are explained using a sample ontology and its instantiation, as depicted in Fig. 6.4. The following namespaces are used in the example code to explain the metrics:

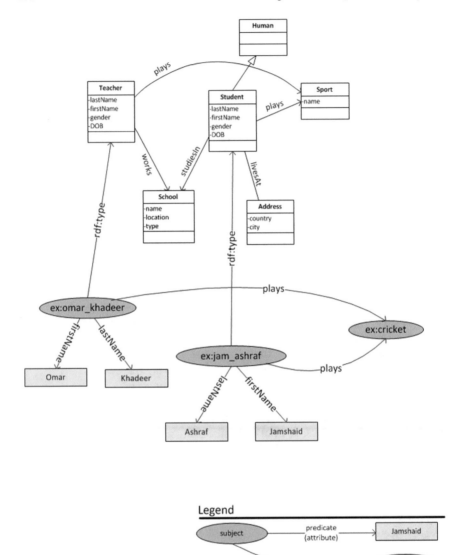

Fig. 6.4 Sample ontology and its instantiation to explain the metrics defined in QUA-AF to measure richness and usage

- **so** is the namespace for the sample ontology. Sample ontology components are referred to using the **so** namespace, such as so: Student
- **ex** is the prefix used to refer to the namespace for instance data such as ex: jam_ashraf

- For W3C-based vocabularies, standard namespaces are used, such as **rdf, rdfs**, and **owl**.

6.3.1 Measuring Ontology Richness

Measuring the richness of ontological terms quantifies the importance of the terms within the ontological model. Ontological terms comprise ontology components such as concepts, object properties (relationships) and data properties (attributes). In the case of RDFS vocabularies, we only consider the object property to refer to the predicates defined by the vocabulary, since object and data properties are not disjoint. The richness of an ontology is measured as *concept richness (CR), relationship value (RV)*, or *attribute value(AV)*. These metrics are explained below.

6.3.1.1 Concept Richness (CR)

Concept richness (CR) defines the structural richness of a concept. When considering a specific concept in an ontology, one needs to consider the relationship it has with other concepts and the number of attributes available to describe its instances. This includes the typed binary relationship (non-hierarchical) with other concepts and data properties providing attribute values for the data description of the concept. Formally, the concept richness of a particular concept $CR(C)$ of a given domain ontology is calculated by adding the number of non-hierarchical relationships and attributes that it has.

$$CR(C) = |P_C| + |A_C| \qquad (6.1)$$

where

P_C *is the number of object, properties (relationship) that Concept C has,* and

A_C *is the number of datatype properties (attributes) that Concept C has.*

A concept's $CR(C)$ value shows its input in representing a specific view of the domain, that is conceptualized by that concept. P_C returns the number of object properties that concept C has, while A_C returns the number of data properties of concept C. The value of $CR(C)$ is a positive integer including zero.

To explain with an example, consider the sample ontology shown in Fig. 6.4. Let C be the *so:Student* concept (i.e. *CR (so:Student)*) and compute the Concept Richness $CR(C)$. The values for P_C and A_C are as follows:

$P_{so:Student} = 3$ because the student concept has three object properties: *so:play, so:livesAt and so:StudiesIn.*

$A_{so:Student} = 4$, because the student concept has four attributes defined for it: *so:lastName, so:firstName, so:gender and so:DOB.*

Therefore,

$$CR(so : Student) = P_{so:Student} + A_{so:Student} = 3 + 4 = 7$$

6.3.1.2   Relationship Value (RV)

The relationship value reflects the possible role of the object property in creating the typed relationships between different concepts. The object property links the instances of the concepts defined as the domain of this property with the instances of the concepts defined as the range of the property. RV is computed as follows:

$$RV(P) = |dom(P)| + |range(P)| \tag{6.2}$$

where

dom(P) is the number of concepts property P has as its domain (rdfs:domain), and

range(P) is the number of concepts (property) P has as its range (rdfs:range).

$RV(P)$ returns an integer, reflecting the number of concepts in which the property can be used to create relationships and provide a rich description of a concept. A property with a higher RV reflects its generalization because more concepts (i.e. instances of the concepts) can be linked through this property. On the other hand, a lower RV value conveys property specificity. Here, a simplified approach is employed to compute relationship richness (RV), since in OWL, the domain and range are taken as axioms and not as type constraints, which could have potentially far reaching effects [225]. Therefore, only the authoritative description[2] of the ontology document to compute RV is referred to, rather than employing the OWL/RDFS model interpretation for domain and range constraints.

Referring to the sample ontology shown in Fig. 6.4, let P be the *plays* relationship. For the computation of relationship value $RV(P)$, compute $dom(so : plays)$ and $range(so : play)$ as below:

$dom(so : plays) = 2$;

because the relationship *so:plays* has two concepts as its domain: *so:Teacher and so:Student.*

$range(so : plays) = 1$;

because the relationship plays has one concept as its range: *so:Sport*

Therefore, $RV(so : plays) = |dom(so : plays)| + |range(so : plays)| = 2 + 1 = 3$

---

[2]The authoritative description of an ontology is the formal ontology document available at the ontology namespace URI [59].

6.3.1.3   Attribute Value (AV)

A concept's attributes are the data properties that provide literal (typed or untyped) values its instances. *AV* reflects the number of concepts that have this data property.

$$AV(A) = |dom(A)| \tag{6.3}$$

Datatype properties are very useful in providing concrete values to describe the concept's instances (individuals). *AV* returns a zero or a positive number, reflecting the number of different concepts using it to semantically describe the concept instances. In the case of RDFS vocabularies, only the *RV* metric to measure the property value is used, since RDFS does not differentiate between object and datatype properties. Referring to the sample ontology shown in Fig. 6.4, let *A* be the *lastName* attribute. For the computation of Attribute Value $AV(A)$, compute $dom(so : lastName)$ as below:

$$dom(so : lastName) = 2;$$

because the attribute *so:lastName* has two concepts as its domain: *so:Teacher and so:Student*.

Therefore,

$$AV(so : lastName) = |dom(so : lastName)| = 2$$

6.3.2   Measuring Ontology Usage

To analyse and quantify the use of ontologies on the Web, the following metrics are defined to incorporate the usage aspect in our framework while analysing its adoption and uptake by publishers. Usage provides an indication of the available instantiation which eventually generates a network effect. Ontology usage is measured as *concept usage (CU), relationship usage (RU)*, or *attribute usage (AU)*. These metrics are explained below.

6.3.2.1   Concept Usage (CU)

The concept usage metric measures the instantiation of the concept in the knowledge base (KB). Here, instantiation means the number of unique URI references used to create members of the class represented by the concept. In the RDF graph, only the triples in which the *rdf:type* predicate is used to create members of a given concept are referred to. The concept usage metric is formalized as follows:

$$CU(C) = |\{t = (s, p, o)|p = rdf : type, o = C\}| \tag{6.4}$$

where

   t is the triple in the dataset describing an instance of type Concept C.

$CU(C)$ returns an integer (including a value of zero), measures the usage of each concept in the knowledge base (dataset), and ranks them based on their instantiation. As mentioned in Chap. 1, however, one of the features of ontologies is the provision of inference, which means using the ontology to classify the data or infer (deduce) new implicit knowledge. This means that by using the axiomatic triples available in the ontology and the reasoning service,[3] additional statements can be accessed which are not explicitly asserted in the knowledge base. This has a direct consequence for the concept usage metrics calculation since $CU(C)$ considers only explicit statements in the dataset. Therefore, an extension to the concept usage metric is proposed to include the subsumption [233] aspect of reasoning by using the entail function.

$$CU_H(C) = |\{t = (s, p, o)|p = rdf : type, o\ entail_{rdfs9}(C)\}| \qquad (6.5)$$

where

   $entail_{rdfs9}(C)$ is a function which implements a reasoning engine based on RDFS [139] and OWL-DL [192] entailment rules. In the concept usage metric, the following RDFS entailment rule $rdf\,9$ is applied:

```
rdf9 :IF(uuu rdfs:subClassOf xxx
         AND vvv rdf:type uuu)
      THEN (vvv rdf:type xxx)
```

This RDF entailment rule will allow top level concepts (super concepts) to subsume the instances of their subclasses.

   Referring to the sample ontology and its instantiation shown in Fig. 6.4, let C be the concept *Student*. The value of metric $CU(so : Student)$ is 1 as there is only one instance of type Student present in the Fig. 6.4. The triple is as follows:

```
<ex:jam_ashraf>   <rdf:type>   <so:Student>
```

RDF data therefore has a usage value of only 1.

6.3.2.2   Relationship Usage (RU)

The relationship usage metric calculates the number of triples in a dataset in which the object property is used to create relationships between different concept instances. From the RDF Graph, the relationship usage value is determined by:

---

[3] Almost all open source and commercial data stores (triple stores) provide RDFS entailment support, such as Virtuoso (www.openlinksw.com) and Stardog (https://www.stardog.com) retr., 12/11/2017).

$$RU(P) = |\{t := (s, p, o)|p = P\}| \tag{6.6}$$

where

  t represents the triples in the dataset having p as their predicate.

The result of RU is a positive integer (including a value of zero). RU with RV helps in indexing the properties which in turn support efficient information retrieval. Referring to the sample ontology and instance data shown in Fig. 6.4, let p be the relationship *plays*. There are two triples in the Fig. 6.4 which have *plays* relationship (predicates). These triples are:

```
<ex:jam_ashraf>   <so:plays>   <ex:cricket>
<ex:omar_khadeer> <so:plays>   <ex: cricket>
```

Thus, $RU(so : plays)$ is computed as follows:

  $RU(so : plays) = 2$

6.3.2.3   Attribute Usage (AU)

The attribute usage metric measures how much data description is available in KB (dataset) for a concept instance. From an RDF graph, AU is calculated as:

$$AU(A) = |\{t := (s, p, o)|p \in A, o \in L\}| \tag{6.7}$$

where

  t is the triple specifying the attribute value for s.

  o in the triple is the datatype property defined in the ontology to provide factual information about the resource being described.

Referring to the sample ontology shown in Fig. 6.4, let A be the *so:firstName* attribute. There are two triples in which *so:firstName* is used to describe the resources. These two triples are:

```
<ex:jam_ashraf>   <so:firstName>   "Jamshaid"
<ex:omar_khadeer> <so:firstName>   "Omar"
```

Thus, the *AU(so:firstName)* is 2.

6.3.3   Measuring Incentive

The incentive metric measures the benefits to the user as a result of using an ontology. The user can be either a data publisher or a data consumer. It hypothesizes the commercial benefits (driving factor) or immediate advantages available in the marketplace (i.e. Semantic Web dataspace) to the users as a result of using the ontology. Given the absence of any statistical data regarding the commercial benefits or advantages available to early adopters of vocabularies in annotating Web information, as mentioned in Sect. 6.2, heuristics based on empirical findings are applied to quantify the incentive.

In the QUA-AF, only data publishers are used when measuring the incentive metric. From the data publishers point of view, the incentive metric determines what benefits the user will obtain as a result of publishing data that is semantically annotated using a particular ontology. For measuring the incentives for data publishers in the QUA-AF, the immediate benefits available to them from search engines as a result of indexing the vocabularies they use to publish the information are analysed. Efforts have been made by several search engines to provide a more powerful search experience for users by including semantically marked data in search results; for example, Yahoo! introduced SearchMonkey in 2008 and a year later, Google announced Rich Snipppet.

To capture such an initiative to measure the incentive for using a particular vocabulary/ontology in the QUA-AF, the direct benefits available to data publishers from search engines as a result of indexing the vocabulary which they use to publish their information are analysed.

Definition (**Incentive of term**). Let $S = \{S_1, S_2 \dots S_n\}$ be the set of search engines[4] which implements the support of a given v (ontology/vocabulary) in their search results.

The incentive value for using an RDF term[5] of an ontology O is calculated as:

$$Incentive_{term} = \frac{1}{n} \sum_{j=1}^{n} w_j * s_j$$

where,

$$\begin{cases} term, & the\ components\ of\ ontology \\ S_j & = 1,\ if\ term\ is\ suported\ by\ search\ engine, otherwise\ 0 \\ n & number\ of\ search\ engines \\ w_j & is\ a\ weight\ factor\ (i.e.\ W_j \in [0, 1])\ and\ the\ sum\ of\ the\ weight \\ & cannot\ be\ more\ than\ 1\ to\ incorporate\ the\ relative\ importance \\ & of\ various\ search\ engines,\ based\ on\ their\ ranking\ approach \\ & and\ also\ country\ coverage. \end{cases}$$

$$(6.8)$$

---

[4]Here only traditional search engines which primarily index non-structured information are considered.

[5]Here *RDF term* refers to the terminological knowledge of an ontology comprising concepts, object properties and datatype properties.

$Incentive_{term} \in [0, 1]$ of a term is a measure of how incentivized the term (concept, property or attribute) is among all the search engines on average. For example, consider the Yahoo and Google indexing services and $t = foaf:surname$ as the term in focus. If it is assumed that term t is recognized by Yahoo but not by Google and each of them is given a weight of 0.5, then the incentive value will be $Incentive_i = 0.5$. The incentive measure can be formulated to include various aspects, such as the number of tools providing building support for the ontology as a whole, or for a certain set of terms of the ontology. We strongly believe that such an incentive serves as a motivating factor for early adopters and helps in bootstrapping the Web of Data on the Web.

6.3.4   Ranking Based on Usage Analysis

The objective of usage analysis is to identify the most highly used terms based on their richness, usage, and available incentives. To rank the terms based on quantitative data, the ranks of a given term t of an ontology O are calculated by aggregating the richness, usage and incentive measures, using their respective metrics. To offer preferential aspects to the ranking, weights are used to adjust the priority of each measure accordingly. To give a consistent representation of each metric, the measure for generating the value in the range of [0,1] is normalized.

The overall rank value of each term is computed as follows:

$$Rank_{t \in O} = W_R\ Richness_t + W_U\ Usage_t + W_I\ Incentive_t \tag{6.9}$$

where

$$if\ t\ is\ concept\ c \in C\ then Richness_{c=t} = \frac{CR(c)}{max(CR(c))} Usage_{c=t} = \frac{(CU_H(c))}{(max(CU_H(c)))}$$

$$if\ t\ is\ relationship\ p \in P\ then Richness_{p=t} = \frac{RV(p)}{max(RV(p))} Usage_{p=t} = \frac{(RU_H(p))}{(max(RU_H(p)))}$$

$$if\ t\ is\ attribute\ a \in A\ then Richness_{a=t} = \frac{AV(a)}{max(AV(a))} Usage_{a=t} = \frac{(AU_H(a))}{(max(AU_H(a)))}$$

and

W_R, W_U and W_I are the corresponding weights of each measure and are adjusted accordingly to the required priority.

Equation 6.9 computes the rank of the terms of the ontology and provides detail on how different measures are computed.

In the next section, a case study is presented which will explain the working of the QUA-AF.

6.4   Case Study: Quantitative Analysis of Domain Ontology Usage

Any typical eCommerce store (website) has numerous web pages pertaining to the various products they offer (product catalogue), promotions (deals), policy-related information, press releases, terms and conditions, warranties, and testimonies from their customers. To improve content accessibility, interoperability and the visibility of their products to a wider group of potential customers (which can be both human and machine agents), the eCommerce store owner would like to annotate the web content using existing ontologies to offer a better means of information dissemination.

To decide which ontologies to use, and which component in those ontologies to use, the eCommerce store owners should:

- identify the ontologies / vocabularies which have some uptake and adoption
- understand their instantiation for a better network effect
- obtain immediate and tangible benefits for semantic annotation offered by search engines

To achieve the above-mentioned requirements, certain tasks which need to be performed in order to understand the importance of ontology usage analysis and its role in implementing such use cases are identified. First, to annotate (semantically describe information) the data, one needs to identify the Web ontologies available for use on the Web. Second, after identifying the available ontologies, one has to understand usage and adoption to identify suitable terms in existing ontologies. The identification of suitable or highly used terms (concepts, properties/attributes) promotes the reusability of existing terms which maximizes the portability of data by consuming applications [140].

In this case study, for succinctness, only those ontologies being used in a domain-focused corpus which is collected by crawling the eCommerce data sources matching the scenario presented in the above use case are identified.

6.4.1   Dataset and Ontology Identification

To conduct an empirical study on the RDF data and analyse the vocabulary usage in a specific (focused) domain, a dataset is built to serve as a representative sample of the Web of Data currently published on the Web. The collected dataset is sufficiently representative to provide a snapshot of actual domain-specific semantic data patterns, enabling meaningful measurements to be made to enhance our understanding of how data is really being used. Using the hybrid crawler and a seed set comprising 259 web domains (web sites), a corpus comprising Semantic Web data is built, as discussed in Sect. 5.6 of previous chapter.

Table 6.1 lists the ontologies found in the dataset used by the data publishers to semantically describe eCommerce-related data. Table 6.1 lists only the ontologies for

Table 6.1 List of ontologies
identified in the dataset

Prefix	Ontology URL
foaf	http://xmlns.com/foaf/0.1/
gr	http://purl.org/goodrelations/v1#
v	http://rdf.data-vocabulary.org/#
dc	http://purl.org/dc/terms/
og	http://opengraphprotocol.org/schema/
rev	http://purl.org/stuff/rev#
vCard	http://www.w3.org/2006/vcard/ns#
virt	http://www.openlinksw.com/schemas/virtrdf
comm	http://purl.org/commerce#
frbr	http://vocab.org/frbr/core#
vso	http://purl.org/vso/ns#
pto	http://www.productontology.org/id

which the authoritative ontology description document from the specified ontology namespace URI were found on the Web. There were some ontologies for which the authoritative description document were not found and these were discarded in this case study. The retrieved ontology documents are stored in the Ontology Library repository to be used by the QUA-AF to perform ontology usage analysis.

In the next section, the developed metrics for the QUA-AF are applied to analyse the usage of different ontologies identified in the dataset. For brevity, the analysis is limited to two largely-used ontologies in the dataset, GoodRelations and FOAF, to enable a detailed discussion on the findings.

6.5 GoodRelations Ontology Usage Analysis

As discussed in the last chapter, the GoodRelations (GR) ontology [142] is an open Web ontology, developed for the eCommerce domain, which allows businesses (and individuals) to describe their *offering, business*, and *products* on the Web. In the version of the authoritative document[6] that was considered in this study, the ontology comprises 32 concepts, 49 object properties and 46 data properties, including a few deprecated terms. From a high level view, the GR model is based on three main concepts, each focusing on a separate aspect of the eCommerce domain. These three concepts are:

- business entity to represent the business organization selling or seeking products;
- offering to represent offers with details of the price; and

---

[6]http://purl.org/goodrelations/v1.owl; retr. 24/11/2017.

- product or service to conceptually describe the product included in the offer made by the business entity.

The QUA-AF is applied on the GR ontology to analyse ontology usage by measuring the concept richness, usage and incentives, as shown in Table 6.2. The table displays the concepts in the order of their final rank value, which is computed using Eq. 6.9.

For the computation of incentives, Eq. 6.8 is used after deciding on the S set. In our approach, three search engines, Google, Yahoo and Bing i.e. $S = \{google, yahoo, bing\}$ are considered the sources which recognize structured data on web pages and particularly meta-data using Web ontologies. It is important to note here that with the emergence of www.Schema.org (which provides the family of schemas to allow web developers to specify structure and unique identifiers for their information, recognizable by the Google, Bing, and Yahoo search engines), the computation of incentives becomes trivial after establishing correspondence between www.Schema.org and the respective ontology.

6.5.1 Computation

The ranking approach allows the specification of the relative importance of each measurement by setting an appropriate weight for richness, usage and incentives at 0.3, 0.5, 0.2, respectively. The numeric values of each measurement are calculated by accessing the knowledge base containing both the terminological statements (T-Box) to measure richness and assertional statements (A-Box) to measure the usage of a given ontology. For example, the CR value of gr:BusinessEntity in Table 6.2 is calculated by querying the ontology graph which returns 5 relationships and 9 attributes giving 14 as the raw value of CR. For CU, SPARQL query (See Fig. 6.5) returns 62,347 instances of business entity type. Given, $t = $ gr:BuisnessEntity is a concept and its incentive value is 0.433, the normalized values with respective weights in Eq. 6.9 is:

$$Rank_t = (0.3 * 14/31) + (0.5 * 62,347/989,638) + (0.2 * .433) = 0.254$$

$Rank_t$ computes the rank value for t. For this experiment, we have given the most weight, comparatively, to the actual usage aspect of the concepts, followed by richness and the lowest weight to the incentives which are adjustable. Moreover, in the computation of incentives, three traditional search engines have been considered: Google, Yahoo and Bing and weights of 0.5, 0.3 *and* 0.2 respectively were given, based on their popularity. The weighting values used are arbitrary and can be changed according to other users' requirements.

Table 6.2 GoodRelations concepts usage analysis and their rank considering richness, usage and incentive measures

Concept Terms	CR	CU	Incentive	Rank
gr:Offering	1	1	0.433	0.887
gr:SomeItems[a]	0.806	0.459	0.167	0.505
gr:ProductOrServiceModel	0.871	0.231	0.233	0.423
gr:UnitPriceSpecification	0.452	0.525	0	0.398
gr:TypeAndQuantityNode	0.194	0.476	0	0.296
gr:ProductOrService	0.710	0	0.233	0.260
gr:BusinessEntity	0.452	0.063	0.433	0.254
gr:Individual[a]	0.774	0.001	0	0.233
gr:QuantitativeValueFloat	0.323	0.243	0	0.218
gr:Location[a]	0.194	0.006	0.333	0.128
gr:DeliveryChargeSpecification	0.419	0	0	0.126
gr:PaymentChargeSpecification	0.387	0	0	0.116
gr:PriceSpecification	0.355	0.001	0	0.107
gr:QuantitativeValueInteger	0.323	0.003	0	0.098
gr:QualitativeValue	0.290	0	0	0.087
gr:QuantitativeValue	0.226	0.035	0	0.085
gr:OpeningHoursSpecification	0.226	0.023	0	0.079
gr:License	0.194	0	0	0.058
gr:DayOfWeek	0.129	0.001	0	0.039
gr:WarrantyPromise	0.129	0.001	0	0.039
gr:PaymentMethod	0.065	0.002	0	0.020
gr:PaymentMethodCreditCard	0.065	0.002	0	0.020
gr:BusinessEntityType	0.065	0.001	0	0.020
gr:BusinessFunction	0.065	0.001	0	0.020
gr:DeliveryMethod	0.065	0.001	0	0.020
gr:DeliveryModeParcelService	0.065	0.001	0	0.020
gr:WarrantyScope	0.065	0.001	0	0.020
gr:Brand	0.065	0	0	0.019

[a]These are the new concepts in the replacement of deprecated concepts. For further details visit http://www.heppnetz.de/ontologies/goodrelations/v1.html; retr 10/12/2017

6.5.2   Observations

In Table 6.2, it can be seen that the highest ranked concept is gr:Offering because it has the highest value of all the three measures; however, there are different concepts which are rich in terms of their description, but due to the usage factor, they have a low ranking score. For example, gr:ProductOrService and gr:Individual are concepts with a high richness value, placing them in 3rd and 4th position in the *CR* index, but due to their lower instantiation (usage), they are placed in 6th

```
1    SELECT ?instance, COUNT(?cls) as ?freq
2    WHERE
3      {
4         ?cls a ?instance .
5         {
6            SELECT ?instance
7            FROM <http://purl.org/goodrelations/v1#>
8            WHERE
9              {
10                  ?instance owl:Class .
11             }
12        }
13     }
```

Fig. 6.5 SPARQL query to compute CU metrics value

and 7th position in the overall ranking. There are six concepts which have no usage (instantiation) in the dataset: `gr:ProductOrService`, `gr:DeliveryChargeSpecification`, `gr:PaymentChargeSpecification`, `gr:QualitativeValue`, `gr:License`, `gr:Brand`. The last two concepts `gr:License` and `gr:Brand` are new concepts recently added to the ontology model, therefore their usage is not evident from the dataset.

Another important observation to make is the *CU* of `gr:ProductOrService` concept is 0, however, it has the 4th highest richness value in the table (see Table 6.2). This is because `gr:ProductOrService` is the super-class of its taxonomical (is-a) hierarchy, having three more specialized concepts, which allows users to annotate data with the most specialized concepts. This use of specialized concepts promotes specificity in describing semantic information, but on the other hand, a user querying the RDF data might use the highest upper-level concept instead. Here, we recall that taxonomic hierarchy implements subsumption behaviours, which in OWL, means necessary implication [225]. This means that all the instances (or individuals) of sub-concepts (sub-classes or leaf-concepts) are also instances of the super concepts (super-class or upper-class). In order to allow the upper-level concepts to reflect the usage of their lower-level concepts, the concept usage metric, which implements the sub-class axioms to subsume the instance of sub-concepts (see Eq. 6.5) was extended. The results shown in Table 6.2 are obtained from the dataset by considering the knowledge available in the dataset and not that which can be inferred using ontology reasoning.

In the incentive column, only very few concepts can be seen to have non-zero values. This is due to the fact that there is very limited evidence available on what is being used from these ontologies by these search engines to index the structured data annotated using explicit semantics.

6.5.2.1 Usage Related Observations

After discussing the approach used to calculate the ranking of ontology terms based on different measures, we present the usage analysis of GR terms. In Table 6.3, the terms are arranged into three groups: concepts, object properties (relationships) and datatype properties (attributes) of the GoodRelations ontology. The rank of each term is calculated by incorporating the three aspects, namely the richness of the term in the ontology; the use of each term in the dataset; and the incentives based on the term's acceptance in traditional search engines. 15 concepts are listed in descending order with `gr:Offering` being the highest ranked in the list, 17 object properties creating relationships between entities, and 17 datatype properties providing textual description to the entities. In the concept list of Table 6.3, it can be seen that `gr:SomeItems`, `gr:ProductOrServiceModel` and `gr:Individual` have a higher ranking than `gr:ProductOrService` which does not even have significant usage, but its richness and incentive values have helped it to rank closely with its specialized concepts. In the object property list, relationships are listed according to their rank value. One can gain a better understanding from the list of how the entities of different types are linked with each other to create a semantic description of the overall eCommerce data. As expected, the properties with highly ranked concepts such as `rdfs:range` or `rdfs:domain` also have a high rank in their listing. For example, the top five properties have `gr:Offering` as their domain with `gr:offers` as range. This helps in realizing the sub-model of the ontology which has high use, forming a light ontology which is useful in such scenarios as data integration and prioritizing the indexing strategy. The third group is a list of attributes with their rank values. These attributes are very useful in exploring the textual description of entities.

The availability of attributes with statistics about their usage is important for querying the data, particularly on the Web where no predefined schema is available (contrary to the relational databases). Knowing which attributes are frequently used allows user interfaces to be built for exploratory search- and knowledge-driven applications. From the datatype property list (Table 6.3), `gr:description` is top of the list, because it is not only heavily used; it also has the highest richness value. This attribute allows the provision of textual information about entities, thus making it highly rich in terms of its coverage and usability. Another notable attribute is `gr:legalName` which, despite having a low richness value, has a high usage value due to its significance in providing human readable names of companies/organizations offering their products on the Web. The terms listed in Table 6.3 enable users to understand the prevalent conceptual schema in a domain-specific implementation (eCommerce, in our use case) and use this information for various application scenarios, including the use case requirement highlighted in Sect. 6.4.

Table 6.3 GoodRelations ontology terms and their ranking

Concept		Object Property		Datatype Property	
Term	Rank	Term	Rank	Term	Rank
Offering	0.887	hasBusiness Function	0.667	description	0.864
SomeItems	0.505	offers	0.575	eligibleRegions	0.433
ProductOr ServiceModel	0.423	availableAt OrFrom	0.566	name	0.406
UnitPrice Specification	0.398	includes	0.554	validFrom	0.246
TypeAnd QuantityNode	0.296	hasPriceS pecification	0.496	validThrough	0.246
ProductOr Service	0.260	typeOfGood	0.393	hasUnit OfMeasurement	0.229
BusinessEntity	0.254	acceptedPayment Methods	0.388	hasStock KeepingUnit	0.169
Individual	0.233	hasManufacturer	0.388	hasCurrency Value	0.126
Quantitative ValueFloat	0.218	eligible CustomerTypes	0.367	hasEAN_UCC-13	0.125
Location	0.128	includes Object	0.345	hasCurrency	0.124
DeliveryCharge Specification	0.126	eligibleTransaction Volume	0.333	valueAdded TaxIncluded	0.092
PaymentCharge Specification	0.116	hasEligible Quantity	0.300	legalName	0.084
Price Specification	0.107	hasMake AndModel	0.300	amountOf ThisGood	0.081
Quantitative ValueInteger	0.098	isAccessoryOr SparePartFor	0.300	hasValue Float	0.058
QualitativeValue	0.087	isConsumableFor	0.300	hasMax CurrencyValue	0.043
		isSimilarTo	0.300	hasMin CurrencyValue	0.043
		hasInventory Level	0.280	hasMinValue	0.037

6.6 FOAF Ontology Usage Analysis

In this section, the FOAF ontology [47], which is regarded as one of the earliest,[7] most highly used and well researched [78, 81, 237] ontologies by the Semantic

---

[7]The FOAF homepage (http://www.foaf-project.org/about; retv 11/11/2017) states that FOAF project started in 2000.

Table 6.4   FOAF ontology terms and their ranking

Concepts		Object Properties		Datatype Properties	
Term	Rank	Term	Rank	Term	Rank
Person	0.606	homepage	0.664	name	0.584
Agent	0.442	Img	0.576	familyName	0.474
Document	0.427	thumbnail	0.554	lastName	0.452
Image	0.339	page	0.471	gender	0.266
Organization	0.301	member	0.406	firstName	0.065
PersonalProfile Document	0.201	maker	0.399	mbox_sha1sum	0.056
OnlineAccount	0.147	isPrimaryTopicOf	0.369	accountName	0.054
Group	0.139	depiction	0.357	status	0.05
OnlineChatAccount	0.119	based_near	0.256	givenName	0.018
Project	0.114	mbox	0.250	title	0.018
		primaryTopic	0.155		
		account	0.147		
		Made	0.147		
		Knows	0.116		
		topic	0.113		
		logo	0.103		

Web community, is examined. In the version used in this study, the FOAF ontology comprises 19 classes, 40 object properties and 27 datatype properties. The FOAF ontology provides the vocabulary to express information about people, their interests, relationships and activities. In Table 6.4, the usage analysis of the FOAF ontology is presented, based on the use of FOAF terms in the dataset.

6.6.1   Observation

The first column in Table 6.4 lists the most highly used concepts in descending rank order. foaf:Person is the mostly instantiated concept used to defined the person entity, followed by foaf:Agent and foaf:Document. It is interesting to note that similar to GR, only a few concepts are used in the implementation of these ontologies on the Web for semantic annotation. In the FOAF ontology, 58.82% of concepts, 40% of object properties and 37% of datatype properties are used with varying frequency, making approximately half of the terms in use and others without instantiation. This usage trend is similar to that in [8, 81], who reported that a small part of the ontologies are, in fact, being used by a large number of data publishers. Such usage patterns are somewhat desirable for promoting a consistent schema to represent entities of interest such as people, places and documents in describing social network information. Referring back to our use case and reflecting on the requirements highlighted under the use case scenario section, the first requirement was to identify the applicable ontologies, which is accomplished by identifying all the ontologies which are presently being used by the relevant community. A domain-specific dataset is used to achieve relevance and specificity. The identified ontologies were then analysed to measure their usage and understand the usage patterns available. The usage analysis helps to identify the terminological knowledge that has better prevalence and prominence in the published data and uses it to construct the web schema to be used for our semantic annotation on the Web.

The next section describes how the results obtained by the QUA-AF can be used to realize the application of OUA.

6.7 Quantitative Analysis Evaluation

The objective of the QUA-AF is to quantitatively analyse the use of ontologies and transfer the analysis into actionable information which can then help to realize the benefits offered by Ontology Usage Analysis. In Sect. 6.4, a use case is presented which uses the QUA-AF to obtain the required insight to implement the use case scenario. Recall that the use case contains the following three main requirements:

- identify the ontologies applicable to the Web site about the eCommerce domain
- understand which terms of a given ontology are highly used and should be reused to achieve a positive network effect
- provide a summarized view of the prevalent use of ontologies in the form of a Web Schema to facilitate the data publishing process, based on the quantitative analysis.

The next subsection discusses how using the results obtained by QUA-AF addresses the above-mentioned requirements of the use case.

6.7.1 Requirement 1: Identify the Ontology's Application to an eCommerce Website

The primary objective of using ontologies and publishing information using ontologies is to enable consuming applications to understand the information in such a way that they can automatically source and link the required information over the the Web. Therefore, it is very important for data publishers to publish information using ontologies which are not only relevant to their application domain but are also recognized by the community. This will help to improve the reusability of ontologies, generating a positive network effect which will enable the benefits of Semantic Web technologies to be realized.

The QUA-AF provides a list of ontologies used in the eCommerce domain which have usage and adoption on the Web. Table 6.1 lists the vocabularies being used covering the different types of information often used to structure eCommerce-related information. For a data publisher or data consumer, it is very useful to know what ontologies are in a given application area. For example, for eCommerce, *foaf, gr, v, dc, og, rev, vCard, virt, comm, frbr, vso, and pto* ontologies are being used as shown in Table 6.1. Each of these vocabularies covers a specific domain, however they are related to each other when co-used to describe information covering different aspects. An eCommerce website needs to describe information about the entities representing the respective domain. These entities include, but are not limited to, "company", "product", "offering", and "location".

6.7.2 Requirement 2: Identify the Ontological Terms to Be Used

After learning which ontologies are being used in a given domain, the next question to arise concerns precisely what is being used from these ontologies. This involves the identification of terms being used and their relationships across different ontologies. The QUA-AF considers three dimensions: ontology richness, ontology usage and commercial incentives, to quantitatively analyse the use of ontologies, which provides the necessary insight to address the above-mentioned requirement. The ranking of terms filters the most used and influential terms that data publishers need to consider when describing data on the Web. For example, in the case of the FOAF ontology, which is often used to describe agents (human and/or non-human) and its attributes, the most used concepts are *Person, Agent, Document, and Image*, as shown in the Concepts column in Table 6.4. Likewise, the object properties with highly ranked values are *homepage, img, thumbnail, page, member and maker* as shown in the Object Properties column in Table 6.4. The five top-ranked data properties are *name, familyName, lastName, gender and firstName* as shown in the Datatype Properties column Table 6.4.

This insight into how the use of ontology-specific terms are quantified using multi-dimensional criteria helps data publishers to understand the present use of a particular ontology from several aspects. In addition to being of assistance to regular data publishers, this information is useful to ontology engineers and domain experts where the availability of such usage-related information provides feedback to inform future thinking. Ontology owners or domain experts can gain a better understanding of the conceptualized model when it has been ranked using multiple aspects. If a concept has a high richness value, which indicates how rich it is in terms of its semantic description and relationship, and also high usage, then it is safe to assume that there is a correlation between the richness of a concept and usage.

6.7.3 Requirement 3: A Summarized View on the Prevalent Use of Ontologies in a Given Application Area

For the GR ontology, there are 15 concepts ranked in descending order with gr:Offering ranked the highest, 17 object properties creating relationships between entities, and 17 datatype properties providing a textual description of the entities, as shown in Table 6.3. The terminological knowledge of the GR ontology which has usage and adoption can be obtained from Table 6.3 for a consolidated view of the given domain ontology. Application developers can use these results to develop prototypical SPARQL queries to retrieve the RDF graph containing the required data elements. Using these statistics, the data publisher can create semantic mapping with other ontologies, knowing which entities can be further described with rich semantics.

Constructing the ontological model based on the usage analysis (i.e. terminological knowledge represented in Table 6.3) will help the ontology developer to extract a light version of the ontology being highly used which could help in ontology evolution. Since the usage analysis can be applied in different scenarios, a threshold value can be given to the value for each metric to obtain a partial list of terms which can be adjusted, depending on the requirements. For example, setting the threshold value of 0.6 for rank would return only one concept i.e. gr:Offering from Table 6.2. If the user is interested in annotating the information to have a better position in the search engine result pages, then the top terms with higher incentive values will be required and, hence, the threshold value can be adjusted accordingly. Likewise, if the user is building a Semantic Web application to consume the data, then using a lower threshold value will assist in querying the dataset to have access to more data. Applying usage analysis on all relevant ontologies will extract the knowledge patterns dominant in each ontology to obtain a meaningful summary of the web schema presently dominating the Semantic Web dataspace.

6.8 Recapitulation

In this chapter, the QUA-AF was presented to quantitatively analyse the use of domain ontologies on the Web. The QUA-AF measures usage from three dimensions: ontology richness, ontology usage and commercial incentives. The inclusion of these dimensions provides erudite insight and a multi-dimensional view of the state of ontology usage and its adoption. Against each dimension, metrics have been developed to measure ontologies from different aspects and a ranking approach is used to obtain a consolidated ranking of the terms of an ontology. A web schema generation use case is used to realise the benefits of the QUA-AF in which, based on the usage of different ontologies in the eCommerce domain, a schema representing the prevalent structure over the Web is generated. The generated Web Schema enables data publishers, data consumers and application developers to understand which ontologies and their components are being used in real world settings. The next phase of the OUSAF is the representation phase in which ontology usage analysis is conceptualized using a formal representation model to allow applications and users to access the analysis results for further processing.

Chapter 7
Representation Phase: Ontology Usage Ontology (U Ontology)

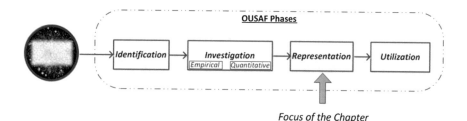

Focus of the Chapter

7.1 Introduction

The huge increase in the use of ontologies and Semantic Web data has increased the need for usage-related information to assist stakeholders (or users) to make effective use of currently available semantic information. As noted in earlier chapters, stakeholders can be different groups of users such as ontology developers, domain experts, application developers and data publishers, each of whom will have a view-specific requirement of the same information. The identification and investigation phases of the OUSAF, discussed in Chaps. 4–6, help us to identify and measure usage-related information. However, once the usage-related information has been determined, it needs to be presented to stakeholders in a structured format; providing granular access to ontology usage-related information therefore meets the needs of each stakeholder. This is done in the *Representation* phase of the OUSAF, in which an ontology usage ontology (the U Ontology) is developed to represent ontology usage analysis-related information. In this chapter, the conceptual framework of the U Ontology is presented.

The chapter is organized as follows. Section 7.2 discusses the various aspects and high level requirements to be considered in representing ontology usage. In Sect. 7.3, a customized methodology and the development phases of the U Ontology are presented. In the subsequent sections, the activities performed in each phase of

© Springer International Publishing AG 2018
J. Ashraf et al., *Measuring and Analysing the Use of Ontologies*, Studies in Computational Intelligence 767, https://doi.org/10.1007/978-3-319-75681-3_7

the methodology are presented. Section 7.4 describes the specification phase, which defines the scope and captures the ontology requirements. Section 7.5 describes the conceptualization phase in which the conceptual model is developed. Section 7.6 describes the formalization phase in which the ontology conceptual model is formally represented. Section 7.7 presents the implementation phase of the adopted methodology which implements the ontology by encoding it, using formal ontology language. Section 7.8 concludes the chapter.

7.2   Different Aspects to be Considered While Representing Ontology Usage

Using the OUSAF, domain ontologies are identified and analysed from multiple aspects in an attempt to establish a detailed understanding of how ontologies are being used, as outlined in Chaps. 4–6. The objective of measuring ontology usage is to provide erudite insight into the usage statistics and usage patterns to facilitate further adoption and uptake. As mentioned previously, this insight influences reusability, evolution and even future thinking on ontology development and reuse. The quantified usage measures obtained need to be made available for consumption and further utilization. Two aspects need to be considered in representing ontology usage-related information: *users* and *structure*. Different types of user are interested in different parts of the information pertaining to ontology usage, therefore their needs should be analysed and considered. Additionally, the structure of information and the mechanism for disseminating usage-related information need to be considered when the representation and utilization phases of the OUSAF are implemented. These two aspects are discussed in the following subsection.

7.2.1   Different Types of User

People often become involved in different stages of the ontology lifecycle and thus need to access information which is specifically relevant to them. Each user may therefore require a different view of the information to perform their desired tasks, based on their role in the ontology lifecycle model. The ontology lifecycle model, as described in Chap. 1, comprises a development stage and in-use stages. The users who interact with ontologies can be categorized as ontology developers/owners, domain experts, application developers and data publishers.

- *Ontology developers* are interested in knowing how the developed ontology is being used and which components of the ontology are either ignored or underexploited. The availability of empirical analysis on the use of ontologies provides the necessary insight for developers to plan changes in their ontologies, and ontology developers are interested to know the usage statistics of the ontology component, including concepts, relationships, attributes and axioms. Identifying the use of concepts that appear in multiple data sources, and their relationships, provides a snapshot of the invariant knowledge patterns of the ontology. Such information is useful to ontology developers in terms of understanding the needs and usage behaviour of users.

- *Application Developers* are interested in knowing what sort of data is available for their applications. Information about the use of ontologies helps developers to anticipate the nature and structure of data, which assists in the development of data-driven applications and interfaces. In the case of linked data-driven applications [159], developers can take advantage of the available terminological knowledge to support development activities.

- *Data publishers* are interested in knowing which ontologies are highly used, and which fragment of any given ontology is the more dominant. One of the immediate benefits which motivates data publishers to publish semantically rich structured data is the availability of machine understandable and processable information on the Web. Prominent search engines like Google (www.google.com) have also started to parse the structured data embedded within Web pages, which motivates publishers to publish their data with them. However, to take advantage of the available benefits, it is important for them to reuse vocabularies which are already used by the community [140].

7.2.2   Structure and Format

The second aspect to consider is the format and structure in which the ontology usage analysis is represented. Ontologies are based on Web architecture [20] and are intended to be equally useful to humans and machines, so information-related ontology usage should preferably be based on these architectural principles. In representing usage analysis, the future possibilities of information processing, the availability of globally accessible data and the definition of canonical terms which can promote information interoperability therefore need to be considered.

Considering the above-mentioned key aspects (users and structure), which also can be seen as non-functional requirements, a domain independent, machine-readable conceptual model in the form of an **Ontology Usage Ontology (U Ontology)** is proposed to represent domain ontology usage. The U Ontology is an ontology which formalizes the representation of ontology usage analysis by standardizing the domain knowledge related to ontology usage analysis. It represents the use of an ontology, the use of its different components, usage statistics, and co-usage with other ontologies.

The development methodology and developmental stages of the U Ontology are presented in the next section.

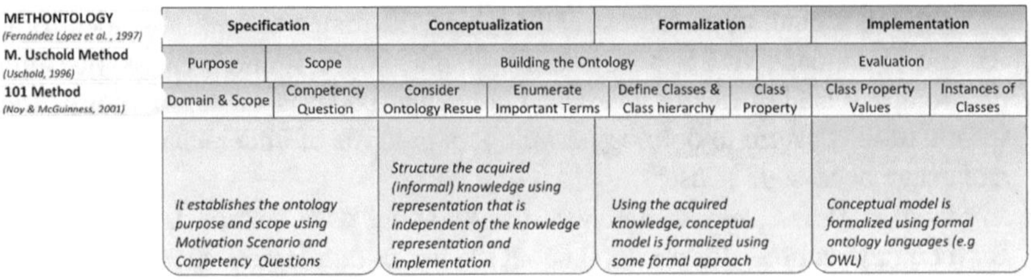

METHONTOLOGY (Fernández López et al., 1997)	Specification		Conceptualization		Formalization		Implementation	
M. Uschold Method (Uschold, 1996)	Purpose	Scope	Building the Ontology				Evaluation	
101 Method (Noy & McGuinness, 2001)	Domain & Scope	Competency Question	Consider Ontology Resue	Enumerate Important Terms	Define Classes & Class hierarchy	Class Property	Class Property Values	Instances of Classes
	It establishes the ontology purpose and scope using Motivation Scenario and Competency Questions		Structure the acquired (informal) knowledge using representation that is independent of the knowledge representation and implementation		Using the acquired knowledge, conceptual model is formalized using some formal approach		Conceptual model is formalized using formal ontology languages (e.g OWL)	

Fig. 7.1 Different ontology development methodologies and their relationship

7.3   Methodology Adopted for U Ontology

As discussed in Chap. 2, several methodologies to support ontology development-related activities have been proposed in the literature. For the development of the U Ontology, different methodologies were studied [56, 97] to identity one that was suitable for U Ontology construction. Three methodologies which were studied in depth are: METHONTOLOGY [98]; Ontology Development 101 [205]; and Mike Uschold [268] as depicted in Fig. 7.1. Four major phases were adopted from METHONTOLOGY for the development of the U Ontology, and the other two methodologies (Ontology Development 101 and Uschold) were used in the realization of these phases. The combination of these methodologies provided the flexibility to adopt those activities most suitable for the development of the U Ontology. METHONTOLOGY [98] is one of the better known methodologies due to its suitability for building ontologies either from scratch or by reusing other ontologies. The methodology comprises four main phases, namely *Specification*, *Conceptualization*, *Formalization* and *Implementation*.

The four phases of the U Ontology development methodology and the set of activities involved in each phase are discussed in the following subsections.

7.3.1   Specification Phase

The aim of the specification phase is to capture and document the ontology requirements. Activities such as capturing the end users' goals, use cases and motivational scenarios demonstrating those goals come in this phase. The Purpose and Scope phases of Uschold's [268] unified approach and the Domain and Scope identification and Competency Question activities of Noy's 101 Method [205] closely relate and overlap with the definition of the specification phase and are used to identify the scope and boundary of domain knowledge. The required formality and format to represent and document the ontology specifications are not specified by any methodology; however, Uschold et al. [268], classify the level of formality of ontologies as: highly formal, semi-informal, semi-formal or rigorously formal. They propose using the motivation scenarios and informal competency questions to specify the scope and capture the requirements of the ontology. In other work, Sure et al. [251], suggest an ontology requirement specification document which describes the following set of information:

- Domain and goal of the ontology.
- Guidelines for designing the concepts, instances and conventions to be followed.
- Scope, including the terms to be represented and the background to capture the prior knowledge.
- Users, use cases and application support for ontology.

As indicated in the discussion above, the specification of an ontology is required to capture the scope of the ontology and elicit the requirements specifications to be used for ontology modelling. Therefore, for the development of U Ontology, the requirements of the specification phase are:

- Capture the scope of the ontology.
- Identify the key users.
- Describe common use case scenarios to spell out the requirements in a more descriptive way.
- From each scenario, extract the key requirements in the form of competency questions to develop the detailed ontology requirements specifications.

7.3.2 Conceptualization Phase

In conceptualization phase, vocabulary which represents the domain needs to be identified and documented to specify the terminology of the domain. The terminology helps in building a common vocabulary within a domain to identify the basic concepts and the relationship between these concepts. This phase in the Uschold et al. [268] methodology is known as Building the Ontology, in which the definition of terms and the ontology itself is built. The following activities are suggested for the conceptualization phase in [98].

- Considering the common vocabulary of the domain, build a complete Glossary of Terms (GT). Since the terms represent the common vocabulary, the glossary contains the concepts, instances, relationships and attribute properties.
- Group the identified terms into concepts and relationships and build a concept classification tree for each group of closely related terms.

These steps are used to produce the conceptual model which can then be used to verify the models usefulness and usability. For the conceptualization phase, Corcho et al., [63] proposed 11 steps to carry out the required activities based on METHONTOLOGY.

The activity proposed by Noy et al. [205] for the ontology reusability stage is to identify the potential ontology candidates for reusability. After the identification of terms and the grouping of terms, it is worth checking to see whether someone has already undertaken conceptualization for a similar domain. Aside from being best practice to reuse existing ontologies as best practice, it is sometimes a requirement if the system needs to interact with other systems that have already committed to a particular ontology.

In light of this, the conceptualization of an ontology is required to capture the key terminology describing the domain knowledge. Therefore, for the development of the U Ontology, the requirements of the conceptualization phase are:

- Identify the key terminologies describing the domain knowledge.
- Based on the identified terminology (vocabulary), search for similar terminologies in existing ontologies.
- Evaluate the potential reusable terms to verify their applicability to U Ontology.
- Structure the key terms based on their relationships.

7.3.3 Formalization Phase

The goal of this phase is to formalize the conceptual model developed in the previous phase (conceptualization). Formalization refers to the creation of a neutral ontology formulation that is independent of the underlying language and platform [129]. The transformation of the conceptual model can be performed at different levels of formalization, ranging from a fully formal model to a semi-computable model, depending on the implementation requirements of the ontology. One of the commonly preferred formalisms is the object-oriented modelling language, due to familiarity with the object-oriented modelling paradigm and the availability of its tools [119]. Methodologies are mainly focused on the conceptualization stage of ontology development [205], therefore this suggests the development of class hierarchies and the specification of values for properties defined in the conceptual model.

Fernandez-Lopez et al. [98] also included integration at this stage to incorporate all the definitions considered for reuse in the formalized model. This aids the identification of inconsistencies which may have occurred as a result of the inclusion of concepts defined in other ontologies, which can then be resolved before the implementation stage.

It is evident from the discussion above that the formalization of an ontology requires the representation of a conceptual model using a formal approach that is independent of the underlying platform and application settings. Therefore, for the development of the U Ontology, formalization requires:

- selecting a formal model for formalizing the U Ontology's conceptual model
- using the selected formal approach to formalize the conceptual model
- integrating the formal model with the other ontological models that are selected for reuse in the conceptualization phase.

7.3.3.1 Implementation Phase

The goal of the implementation phase is to encode the formalized model using a formal ontology language. Usually, implementation tools such as ontology editors are used for ontology implementation. Several ontology development environments are available, such as: Protg [168], NeOn Toolkit [133], TopQuadrants TopBraid Composer [261] and OntoEdit [251]. Ontology languages with varying levels of expressivity are available [114], depending on expressivity requirements. However, the W3C-based ontology languages such as RDFS and OWL (including its different species such as OWL-Lite, OWL-DL and OWL-Full) are commonly used to encode ontologies in a formal ontology language.

The implementation of an ontology requires the encoding of the formal conceptualized model. Therefore, for the development of the U Ontology, the requirements of the implementation phase are:

- choosing the formal language to be used to encode the ontology
- selecting the ontology development environment to support the ontology construction-related activities.

Using the adopted methodology, the steps in the development of the U Ontology are as shown below.

7.3.4 Steps Involved in the Development of U Ontology

For the development of the U Ontology, a customized approach is adopted. Methods utilized for specification, conceptualization, formalization and implementation are shown in Fig. 7.2 and discussed in the following subsections.

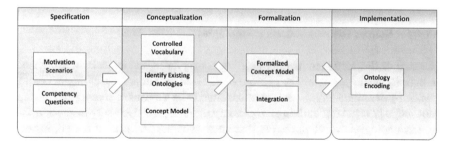

Fig. 7.2 Customized ontology development methodology for the U ontology

7.3.4.1   Step 1: Develop Motivation Scenario

The goal of this step is to develop detailed motivation scenarios. The purpose of motivation scenarios is to obtain a clear picture of the scope of the ontology [268]. Another advantage of developing motivation scenarios is that they assist with gleaning the concepts, terms and relationships that will help to develop the controlled vocabulary and conceptual model. As suggested by [268], the motivational scenario should start with a general scenario and evolve into specific scenarios in a hierarchical manner.

7.3.4.2   Step 2: Develop Competency Questions

The goal of this step is to develop competency questions based on the developed motivation scenarios. The purpose of competency questions is to express the requirements and usage scenarios in a detailed and thorough manner to assist in capturing the complete domain knowledge expected to be represented by the ontology. Competency questions also help to confirm if the ontology adheres to the developed use cases.

7.3.4.3   Step 3: Develop Controlled Vocabulary

The goal of this step is to develop a glossary of terms that represent the domain knowledge. The purpose of controlled vocabularies is to maintain a complete list of terms being used by users and domain experts during discourse. The glossary of terms helps the classification of different components of speech, such as nouns and verbs, and identifies the entities being represented by the domain. As suggested by Fernandez-Lopez et al. [98], if motivation scenarios and their competency questions are documented well, they can become the main source of input for building domain specific vocabularies.

7.3.4.4   Step 4: Identify Existing Ontologies

The goal of this step is to identify existing vocabularies and consider reusing their definitions where possible. The purpose of this is to encourage reusability in ontologies, as by definition, ontologies are understood to be a means of sharing and reusing knowledge [235]. After building the glossary of terms, it is necessary to check existing terminologies (vocabularies) to determine whether any can be reused, instead of developing the required terms from scratch. To do this, one can access ontology libraries and/or undertake a web search to find ontologies which have similar definitions of terms. Manual effort is required to find the similarity between these terms and decide which terms to reuse or develop from scratch.

7.3.4.5   Step 5: Develop Conceptual Model

The goal of this step is to develop the conceptual model of the domain in focus. The purpose of the conceptual model is to arrange the identified concepts (terms) in structural and hierarchical order to group the relevant terms. The conceptual model remains the same, regardless of the formal model and formal language used later to model and implement the ontology. The conceptual model encompasses five components, namely the classes, relationships, functions, instances and axioms in it along with their constraints. Relationships, particularly taxonomical relationships, are included in the model to reflect the is-a relationships present in different terms (concepts).

7.3.4.6   Step 6: Formalize Conceptual Model

The goal of this step is to formalize the conceptual model. The purpose of formalization is to create a neutral ontology formulation that is independent of the underlying implementation language that can be used to serialize the ontology [129]. One of the preferred choices for formalizing the conceptual model [65] is the use of the Unified Modelling Language (UML). UML is an industry-based standard modelling language which provides graphical notation, a set of diagrams and other components necessary for developing software engineering models. A UML Class diagram is used to represent the concepts of ontologies and comprises three elements: the name of class, attributes of class and operations of class, as shown in Fig. 7.3a. In the context of ontology modelling, classes are known as concepts and attributes are known as attributes of concepts; however, operations of classes are not required because ontologies do not have operations. The modified notational diagram to represent the concept of an ontology is shown in Fig. 7.3b.

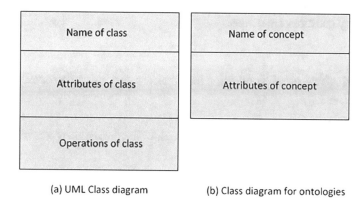

(a) UML Class diagram   (b) Class diagram for ontologies

Fig. 7.3   UML class diagram for ontology modelling

7.3.4.7   Step 7: Integrate with Other Ontologies

The goal of this step is to integrate the formalized conceptual model with the external ontologies considered for reusability. The purpose of integration is to verify that the set of new, reused definitions are aligned and based on the set of new basic terms. This will determine whether any inconsistencies exist prior to implementation. It involves verifying the semantic similarity between terms that are syntactically similar but that may not necessarily be semantically similar.

7.3.4.8   Step 8: Ontology Encoding (Implementation)

The goal of this step is to codify the conceptual model using a formal ontology language that is understandable by the different types of users. An ontology development environment such as Protg [168] is used to develop the ontology, and syntactic and lexical analysers assist in resolving the syntactical and lexical errors [98]. A modern development environment such as Protg allows the user to plug in semantic (OWL) reasoners [33] to help users to avoid inconsistencies which may arise in the development of the ontology model or during the integration with external ontologies. Protg is used as the development environment for the development of the U Ontology and encoding is based on OWL-DL expressivity.

In the next section, each of these steps is discussed in detail along with the implementation details.

7.4   Specification phase: Motivation Scenarios and Competency Questions

In this section, the motivation scenarios are presented and then, using these scenarios, competency questions are defined.

7.4.1   Motivational Scenarios to Capture the Requirements of Users

Four motivational scenarios are presented to describe the requirements of ontology usage analysis from each users perspective.

7.4.1.1   Scenario 1: Ontology Developers (owners)

Ontology owners/developers are usually interested in addressing the following questions:

1. Which components of the ontology are being used and what is the level of usage? This may include the use of different concepts and relationships to interlink concepts.
2. Which attributes are being used by ontology users to provide the instance data describing the entities defined by these concepts?
3. What other terms (vocabularies) are most frequently used alongside the ontology? This will assist in determining the coverage of the instance data being described using the ontologies, and in evaluating the scope of the ontologies, and can be considered as input for ontology evolution.
4. Which components of the ontology have good adoption and which components are marginally used? This information can be obtained via the feedback loop (as depicted in Fig. 1.10) which is based on the actual ontology instantiation.
5. In which application areas is the ontology being used, for example, semantic annotation, data integration, or building ontology-based knowledge applications?

7.4.1.2   Scenario 2: Application Developers

Application developers are usually interested in knowing the availability of different ontologies and how their components are used in a given domain to use the available information in a more effective manner, hence, they would be interested in the following issues:

1. What is being used and what is its adoption level in a given ontology? The availability of this information helps to make effective use of Semantic Web data.
2. How are the different ontologies linked? Information about ontology usage and co-usage with other ontologies helps to obtain a snapshot of the data structure of the published data to aid the development of data-driven applications that are primarily based on data published on the Web.
3. How are different textual properties (also referred to as labelling properties) used? This information is useful for developing user interfaces for data-driven applications.
4. What are the statistics on the use of different concepts, properties and attributes? This information will assist in anticipating the data load and planning data management.

7.4.1.3 Scenario 3: Data Publishers

It is very important for data publishers to know about and learn ontologies that have already been adopted, because their reuse offers better value for the publishing effort. Data publishers are usually interested in the following information:

1. Which ontologies are being frequently used and what is their level of adoption? This information will help in semantically annotating application area-specific information on the Web. Increasingly, search engines are recognizing structured data embedded in Web pages which has been annotated using ontologies, therefore using ontologies that are already supported is highly desirable for data publishers to increase the visibility of their data. Reusing already used concepts which reflect community consensus generates a positive network effect, increasing data visibility and value.
2. What are the usage statistics? This information will assist in quantifying usage in order to decide which ontology or ontology components should be used to achieve the desired objectives.

7.4.1.4 Scenario 4: Semantic Web practitioners/users

Semantic Web researchers/users are usually interested in the following information:

1. What are the prevalent knowledge patterns on the Web? This information helps Semantic Web users and researchers to understand the prevalent structure of information invariantly published on the Web, which will assist them to analyse and infer the relationships between the ontology conceptual model and the data structure prominent on the Web.
2. Which data patterns which are semantically annotated using domain ontologies that are available on the Web?
3. How are different ontologies used by data publishers to semantically describe an entity?

The insight gained from this information will help ontology engineers and Semantic Web users to understand the common needs of data publishers, which can influence future thinking and research agendas.

To summarize, in order to address the requirements specified in the four motivating scenarios, a detailed and multi-dimensional analysis of ontology usage is needed. Aside from identifying this information, a mechanism is required to represent the information (pertaining to ontology usage) in such a way that it can be accessed programmatically for automatic processing. The high level requirements that need to be identified and represented can be summarized as follows:

- Obtain the list of ontologies that are being used in a given application area
- Obtain an analysis that covers different aspects of ontology usage
- Obtain a list of ontologies to semantically describe the entities of the specific domain

- Identify prevalent knowledge and data patterns
- Obtain the usage statistics of ontology components such as concepts, relationships, attributes and axioms.

The acquisition of this information will help to identify the scope of the U Ontology which conceptualizes the domain of ontology usage and its analysis.

7.4.2   Competency Questions (CQ) to Capture the Scope of Representation

Based on the four motivation scenarios and the detail required to perform ontology usage analysis empirically and quantitatively, the competency questions (e.g. CQ1) and the sub-questions under them (e.g. CQ1.1) are listed below to specify the precise requirements for the U Ontology.

CQ1   What are the ontologies being used in a given application area (domain)?

    CQ1.1   What are the namespaces of the ontologies being used in a given application area?
        CQ1.1.1   What is the namespace of a given ontology?
        CQ1.1.2   What is the prefix used for a given ontology?
    CQ1.2   What are the components of a given ontology?
        CQ1.2.1   How many classes does an ontology have?
        CQ1.2.2   How many relationships does an ontology have?
        CQ1.2.3   How many attributes does an ontology have?
        CQ1.2.4   How are different axioms being used in a given ontology?
    CQ1.3   How is a given ontology's conceptual model structured?
        CQ1.3.1   How many relationships does a concept have?
        CQ1.3.2   What are the relationships a concept has?
        CQ1.3.3   How many attributes does a given concept have?
        CQ1.3.4   What are the attributes of a given concept?
    CQ1.4   How are the relationships of a given ontology structured?
        CQ1.4.1   How many concepts are in the domain of a given relationship?
        CQ1.4.2   How many concepts are in the range of a given ontology?

CQ2   What is the richness of a given ontology?

    CQ2.1   What is the richness value of a concept?
    CQ2.2   What is the richness value of a relationship?
    CQ2.3   What is the richness value of an attribute?

CQ3   How is a given concept being used in real world implementation?

    CQ3.1   What is the instantiation of a given concept?
    CQ3.2   How are the entities of a given concept type semantically described?
    CQ3.3   What relationships from the ontology are used to describe the entities?

CQ3.4   What attributes are used to provide the factual instance data describing entities?

CQ3.5   What are the concepts of other ontologies of which the given subject (entity) is an instance?

CQ3.6   What other ontologies are being used together to describe the entity?

CQ4   How are the textual descriptions attached to the entities?

CQ4.1   What are the W3C-based vocabulary label properties that are being used?

CQ4.2   What is the usage of these W3C-based vocabulary label properties?

CQ4.3   What other domain-specific labelling properties are being used?

CQ4.4   What is usage of these domain-specific label properties?

CQ5   List the terms of a given ontology which are recognized by the search engines?

CQ5.1   Is the given concept being recognized/supported by the X number of search engines?

CQ5.2   Are the given relationships being recognized/supported by the X number of search engines?

CQ5.3   Is the given attribute being recognized/supported by the X number of search engines?

CQ6   What are the common knowledge patterns in the implementation of a given ontology?

CQ6.1   What is the maximum path length in the traversal path (knowledge pattern)?

CQ6.2   How many unique paths lead from an entity?

CQ6.3   What are the path steps in the traversal path?

CQ6.4   What is the frequency of a given path step?

CQ7   Which ontology components have limited or no usage?

CQ7.1   What concepts have not been used by data publishers?

CQ7.2   What relationships have not been used by data publishers?

CQ7.3   What attributes have not been used by data publishers?

CQ7.4   How can the ontology components that have usage based on the specified threshold value be accessed?

7.5   Conceptualization Phase: Controlled Vocabulary, Existing Ontologies and Conceptual Model

In this section, the conceptualization phase of the adopted methodology is presented. This phase involves the identification of the terminological knowledge describing the domain, identification of existing ontologies for reuse, and development of the conceptual model. Each of these activities is discussed in the following subsections.

7.5.1 Controlled Vocabulary to Identify the Terminological Knowledge

In the conceptualization phase, all relevant terms of an ontology are defined to obtain a controlled vocabulary, as previously noted in Sect. 7.3. The concepts related to ontology usage identified in Chaps. 4–6 and the motivation scenarios and competency questions presented in Sect. 7.4 provide the basis for building the controlled vocabulary for the Ontology Usage Analysis domain. The controlled vocabulary thus developed is presented in Table 7.1.

The next activity is the identification of existing ontologies which can be reused for the development of the U Ontology.

7.5.2 Identify Existing Ontologies for Reuse

The next step is to evaluate existing ontologies to identify the ontologies which have potentially reusable classes and properties

The identified domain knowledge can be easily clustered into three groups of relevant information related to ontology usage analysis: *Ontology Usage*, *Ontology Metadata* and *Ontology Application*, as depicted in Fig. 7.4. The Ontology Usage cluster represents the domain knowledge specific to the usage analysis aspect of the ontologies; the Ontology Metadata cluster represents the domain knowledge specific to the metadata of the ontology; and the Ontology application cluster represents the domain knowledge specific to the application areas in which the ontologies are being deployed.

The Semantic Web community is working on the development of meta-level ontologies to capture the metadata which can be used by other applications or ontologies to access ontology-related information. One such effort is the development of the Ontology Metadata Vocabulary (OMV)[1] [137] which is a standard proposal for describing ontologies and related entities. Members of the ontology community gather annually at the Ontology Summit and publish the summit proceedings on the ONTOLOG website.[2] At the 2011 Ontology Summit,[3] the Ontology Application Framework (OAF) was presented with the aim of defining common terminology to describe applications of ontologies and the benefits that ontologies deliver within these applications.

---

[1]The project details can be found at http://mayor2.dia.fi.upm.es/oeg-upm/index.php/en/technologies/75-omv; retr. 25/11/2017.

[2]http://ontolog.cim3.net/; as part of retr 13/11/2017.

[3]The summit was chaired by Professor Michael Gruninger (University of Toronto) and Dr. Michael Uschold (Semantic Arts), and the Ontology Application Framework was presented. http://ontolog.cim3.net/cgi-bin/wiki.pl?OntologySummit2011; retr. 15/11/2017.

Table 7.1 Controlled vocabulary for ontology usage ontology (U ontology)

Term	Ontology component	Description
AtrributeUsage	Concept	Quantifies the attribute usage
Attribute	Concept	The attributes that are used for the concept being analysed. Attributes are the datatype properties used to provide literal (static) value
AttributeValue	Concept	The richness value computed for a given attribute
ConceptRichness	Concept	The richness value of a concept
ConceptUsage	Concept	The usage of a given concept
Dataset	Concept	The dataset used to measure ontology usage
DataSource		The dataset used to measure ontology usage
DomainLabel	Concept	The use of data properties as defined in the domain ontology to provide the textual description for entities
FormalLabel	Concept	The use of data properties as defined in the domain ontology to provide the textual description for entities
IncentiveDim	Concept	The use of the incentive dimension to measure the commercial advantages available to data publishers
KnowledgePattern	Concept	The knowledge pattern present in the dataset. Knowledge patterns include the concepts and relationships used to describe the information
LabelUsage	Concept	The frequency with which a label is used. This quantifies the use of a label property
Measure	Concept	The measures that are used for Ontology Usage Analysis, containing all the dimensions for which metrics are defined
OntologyUsage	Concept	The use of ontology. This is a high level concept which represents other sub-concepts related to ontology usage and usage analysis
OntologyUsageAnalysis	Concept	Top level concepts to represent the ontology usage analysis domain
Path	Concept	The unique knowledge pattern in the dataset. A knowledge pattern is a set of triples chained together to form a path
PathConcept	Concept	The concepts in the knowledge patterns. Knowledge patterns comprise path steps which are in the form of triples. PathConcept represents the type of subject and object resources

(continued)

Table 7.1   (continued)

Type	Ontology component	Description
PathProperty	Concept	The relationship in the knowledge patterns. Knowledge patterns comprise path steps which are in the form of triples. PathProperty represents the predicate of the triple
PathStep	Concept	The data or knowledge patterns. A path step represents a triple present in the dataset
Relationship	Concept	The use of object properties to describe the instance of Concept (i.e. ConceptUsage)
RelationshipUsage	Concept	The use of relationships (object properties) of the ontology
RelationshipValue	Concept	The computed richness value of a relationship
RichnessDim	Concept	The richness dimension to measure the structural characteristics of the ontology
SearchEngine	Concept	The search engine that supports the particular term (concept, relationship or attribute)
SoftwareSupport	Concept	The software component, which could be an application, API, database or Reasoner that support the particular term (concept, relationship or attribute)
Source	Concept	The source that is being used as input for analysing ontology usage. This is a high level concept
Streaming	Concept	The data source which is accessed by continuously crawling the Web
Term	Concept	A high level concept representing the terminological knowledge of the ontology
UsageDim	Concept	The usage dimension to measure the use of an ontology and its components in real world implementation
Vocab	Concept	The different vocabularies/ontologies that are used to describe the instance of a concept
analysesOntology	Relationship	The ontology that is being analysed by OUA
attributeValue	Relationship	The richness value of an attribute
hasAttribute	Relationship	The attributes that are used to provide attribute values to the entity defined by the concept
hasAttributeUsage	Relationship	The use of attributes (data type properties) of the ontology that is being analysed

(continued)

Table 7.1 (continued)

Type	Ontology component	Description
hasConceptUsage	Relationship	The concept of the ontology whose usage is measured and analysed
hasDomainLabel	Relationship	The domain label used for a concept
hasFormalLabel	Relationship	The formal label used for a concept
hasIncentiveDim	Relationship	The incentive (commercial benefits) dimension of the ontology usage
hasKnowledgePattern	Relationship	The knowledge patterns of a given ontology
hasLabelUsage	Relationship	The use of label properties. The label properties are used to attach textual descriptions to entities
hasMeasure	Relationship	The measures used to perform ontology usage analysis
hasObjectInPath	Relationship	The object of the triple represented by the path step
hasPath	Relationship	The specification of the paths present in the knowledge patterns
hasPathStep	Relationship	The specification of the path steps included in the path
hasPropertyInPath	Relationship	The predicate of the triple represented by the path step
hasRelationship	Relationship	The relationships (use of object properties) that are used to describe the entity defined by the concept
hasRelationshipUsage	Relationship	The use of relationships (object type properties) of the ontology that is being analysed
hasRichness	Relationship	The richness value of the concept
hasRichnessDim	Relationship	The richness dimension of the ontology usage
hasSubjectInPath	Relationship	The subject of the triple represented by the path step
hasTerm	Relationship	The terminological knowledge of the ontology that is being analysed: concepts, relationships (object property) and attributes
hasUsageDim	Relationship	The usage dimension of the ontology usage
hasVocab	Relationship	The specification of the vocabularies that are being used in order to describe the entity of a concept
isComponentOf	Relationship	The association of a term with its ontology

(continued)

Table 7.1   (continued)

Type	Ontology component	Description
isCoused	Relationship	The two ontologies which are being co-used in the dataset
isIncentivizedBy	Relationship	The relationship between the ontology term and the term supported by the search engine
isSupportedBy	Relationship	The relationship between the ontology term and the software that supports the term
isUsedBy	Relationship	This property allows the data sources (website/pay-level-domain) included in the dataset to be specified. It refers to data sources which make use of ontologies to describe information on the Web
performedOn	Relationship	The specification of the source which is used to perform the ontology usage analysis
relationshipValue	Relationship	The richness value of a relationship
analysisTimestamp	Attribute	The date and time the reported usage analysis was performed
description	Attribute	Textual description of the resources
docURI	Attribute	The URL (internet address) hosting the ontology related documents
frequency	Attribute	The occurrence of the term
hasCousedValue	Attribute	This represents the number of other ontologies being co-used with the given ontology The value is represents the co-usage value obtained through the Co-Usage ontology network (projected network degree)
hasInstantiation	Attribute	The number of instances of a concept
hasUsers	Attribute	This property allows the total number of users a particular ontology has to be specified. It represents a numeric value to tell how many different data sources are using the ontology
incentiveValue	Attribute	This represents the incentive value for a term computed by the incentive metric
industry	Attribute	This property reflects the domain to which the crawled data belongs. It is a manual entry to classify the industry to which apparently crawled data belongs
name	Attribute	The name of the concept and other resources
prefix	Attribute	This is the prefix used in the dataset and during the analysis to refer to ontology

(continued)

Table 7.1 (continued)

Type	Ontology component	Description
searchEngineName	Attribute	Specifies the name of the search engines considered to measure the incentive value of each supported term
timestamp	Attribute	The date and time the information is obtained
URI	Attribute	The URI of the ontology and ontology components
URL	Attribute	This represents the address of the data source from which the Semantic Web data is crawled
usageValue	Attribute	This is the usage value of a term (concept, relationship, attribute) computed by the Ontology Usage metric
value	Attribute	The numeric value

Fig. 7.4 Ontology-related information clusters

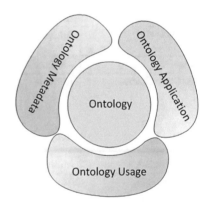

These two projects, which are described in the following subsections, complement the ontology usage analysis domain and relate to the Ontology Metadata cluster and the Ontology Application cluster, shown in Fig. 7.4.

7.5.2.1   Ontology Metadata Vocabulary

The objective of the Ontology Metadata Vocabulary (OMV) is to provide a standard approach to describe ontologies and related entities [137]. The OMV ontology is considered an ontology metadata standard for annotating ontologies. The use of OMV promotes the features which enhance reusability that are equally accessible for both humans and machines. OMV is designed modularly and comprises OMV code and OMV extensions. The OMV code captures the key information that is relevant to the majority of ontologies, whereas the OMV extension allows ontology users to provide more specialized, application-specific, ontology-related information.

There are two main classes in OMV around which other concepts are defined. These are `OntologyDocument` and `OntologyBase`. `OntologyBase` (OB) represents the conceptualization of the ontology, whereas `OntologyDocument` (OD) represents the realization of the conceptualized ontology. The other classes in OMV core are Person and Organization to specify the party responsible for creating, reviewing, contributing and applying the ontology. The method, tools and formal language used for developing the ontology are described using `Ontology EngineeringMethodology`, `OntologyEngineeringTool`, and `OntologyLanguage` classes, respectively. The serialized form of an ontology is available at http://omv2.sourceforge.net/ (retr., 09/11/2017) and for more descriptive details, readers are referred to [216].

7.5.2.2 Ontology Application Framework

As previously mentioned, the Ontology Application Framework (OAF) was presented at the Ontology Summit 2011 to make a case for the use of ontologies by providing concrete application examples, success/value metrics and advocacy strategies [269]. The objective of OAF is to present a common terminology for describing the application scenarios in which ontologies are used. It also captures the benefits and value that can be achieved from the applications as a result of the use of ontologies. Additionally, it provides a basic vocabulary to represent benchmarks and has the ability to compare different applications of ontologies [271]. Several of the areas of ontology use are: integration, decision support, semantic augmentation and knowledge management. The conceptual representation of OAF is available at http://ontolog.cim3.net/file/work/OntologySummit2011/ApplicationFramework/OWL-Ontology/OntologyApplicationFramework-WithDocumentation.pdf (retr. 19/10/2017).

7.5.3 Conceptual Model for U Ontology

In the conceptual model, the key concepts of the domain that have been previously identified are structured to show their relationships with each other and to specify restrictions in their relationships. As mentioned in [131], the structural representation and its components remain independent regardless of the formalization language and approach used to serialize the conceptual model. Therefore, by using the terminology introduced in Table 7.1 and grouping the concepts that are related to each other, the following conceptual sub-models are presented.

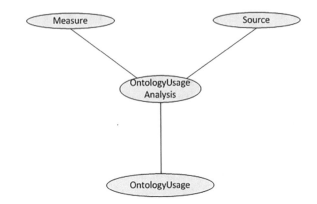

Fig. 7.5 Ontology usage analysis sub (conceptual) model

7.5.3.1   Ontology Usage Analysis Sub-model

The sub-model shown in Fig. 7.5 relates the core concepts of the ontology usage analysis domain model. `OntologyUsageAnalysis` is a high level concept that represents the ontology usage analysis domain and its activities. As shown in Fig. 7.5, it comprises three main components `Source`, `Measure` and `OntologyUsage` which are explained in the following subsections:

Source: Source refers to the data source that is being used by OUA to obtain the required inputs for its analysis. Dataset and Streaming are the two specialized data sources utilized as inputs in the conceptual model. Dataset is the repository which holds the crawled Semantic Web data and Streaming represents the continuous crawled data which is input to the OUA through RDF data stream processing [188].

Measure: Measure represents the different dimensions from which the use of ontologies are measured. Measure is a concept which can generalize any of the dimensions that are applicable to ontology usage analysis. In this model, three dimensions are used to analyse ontology usage and their corresponding concepts are `UsageDim` (defined in Sects. 5.4.3 and 6.3.2), `RichnessDim` (defined in Sect. 6.3.1), and `IncentiveDim` (defined in Sect. 6.3.3). `UsageDim` represents the "usage" dimension in which ontology usage is measured; `RichnessDim` represents "richness" which measures the ontology component's structural characteristics; and `IncentiveDim` represents the incentive dimension which captures the commercial advantages available to data publishers as a result of the use of ontologies.

OntologyUsage: This is the central or pivotal concept of the ontology usage analysis domain. As indicated by its name, it represents the overall discipline in which ontologies and their components are analysed. `OntologyUsage` also represents the usage analysis of ontology components. It conceptualizes usage analysis for each component through specialized concepts, as shown in Fig. 7.6. `ConceptUsage` represents the different aspects from which a concept (here concept represents the class of ontology that is being analysed) of a given domain ontology is analysed. The `ConceptUsage` sub-model is discussed in the next subsection.

Fig. 7.6 OntologyUsage
and related concepts

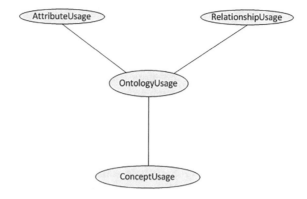

RelationshipUsage represents the usage analysis of the object properties and similarly, AttributeUsage represents the use of attributes defined in the domain ontology.

7.5.3.2 Concept Usage Sub-model

ConceptUsage (previously defined in Sect. 5.4.3) is the concept which represents all the aspects from which a concept is being analysed. The ConceptUsage sub-model, depicted in Fig. 7.7, shows the four aspects namely: Relationship, Attribute and LableUsage that are analysed for a given concept. Each of these aspects is discussed below.

   Vocab: Vocab represents the different vocabularies used to describe the entity (the instance of the concept). As a commonly required and recommended best practice, the entities (concept's instance) are described by using the relationships that are defined by the domain ontology and also the other ontologies/vocabularies that are common in the respective community. Therefore, to establish a comprehensive

Fig. 7.7 Concept usage
sub-model

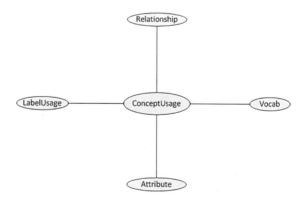

understanding, it is important to know which vocabularies/ontologies are being used by data publishers to semantically describe the entities. `Vocab` captures all such ontologies whose terms are used to describe the resource.

Relationship: `Relationship` captures the use of different object properties to semantically describe the entities. The relationships used can come from the domain ontology that is being analysed or from other ontologies, therefore it is important to learn about all the object properties that are used by different data publishers, because this covers the entity relationships with other entities.

Attribute: `Attribute` captures all the datatype properties that are used to provide the attribute description of the entities. These are normally literal values which provide factual statements about entities. Therefore, it represents all the data properties used by data publishers to provide factual statements describing the state of entity.

LabelUsage: `LabelUsage` (defined in Sect. 5.4.4) captures the use of label properties that provide the textual description of the entities. Label properties are normally used to provide human-friendly information about entities. As defined in Chap. 5, in the OUSAF, label properties are categorized into two types: Domain Labels (DL) and Formal Labels (FL). These are considered to be the standard labelling properties for providing textual information; `DomainLabel` and `FormalLabel` are therefore the specialized concepts of `LableUsage` for capturing domain labels and formal labels, respectively.

7.5.3.3   KnowledgePattern Sub-Model

`KnowledgePattern` (defined in Sect. 5.4.5) comprise `Path` which has `PathStep` to represent each triple in the path. `PathStep`, which represents a single triple, comprises `PathConcept` and `PathProperty` to specify the concepts and predicates respectively of the triple, as shown in Fig. 7.8. These terms are described as follows:

PathConcept: `PathConcept` represents the concepts of the entities that are present in a triple. This includes the concept used as both a subject and an object.

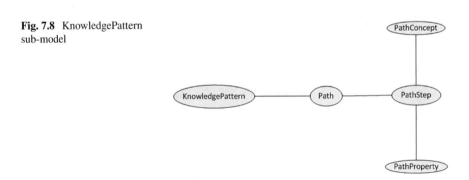

Fig. 7.8 KnowledgePattern sub-model

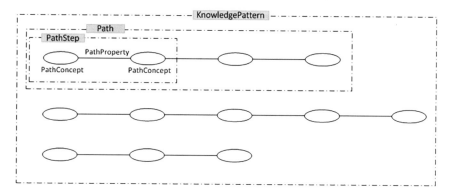

Fig. 7.9 KnowledgePattern components

PathConcept is shown in Fig. 7.9, which depicts the anatomy of the knowledge pattern. PathConcept captures the concept of the ontology that is being used to instantiate the subject and the object of the triple described using PathProperty.

PathProperty: PathProperty represents the predicate present in the triple which forms the PathStep. PathProperty captures the object property that describes the subject by creating a typed relationship with another resource, as shown in Fig. 7.9.

PathStep: PathStep represents the single triple that is included in the Path. PathStep is shown in Fig. 7.9 by a dotted line containing the triple inside it.

Path: Path represents the unique sequence of triples (PathStep) linked together to describe a portion of the domain knowledge. Path is shown in Fig. 7.9 by a dotted line which contains several path steps.

7.6 Formalization Phase: Ontology Formalization and Integration

The formalization phase is the third phase in the development of the U Ontology in which the conceptual model is formalized using a formal modelling approach. This phase contains two sets of activities: formalization of the conceptual model and integration with existing ontologies (for reusability). These two activities are described in the following subsections.

7.6.1 Formalization of Conceptual Model

UML is considered to be an industry standard for modelling a conceptual model and provides a set of graphical notations to describe the model components. A class

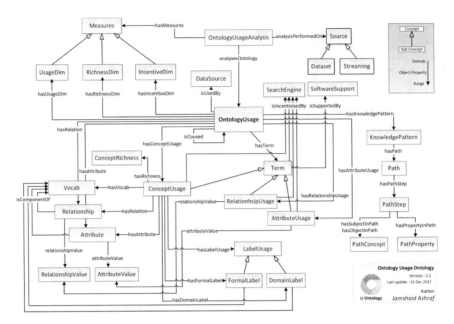

Fig. 7.10 U ontology overview (V2.2)

diagram is often used to formally represent the ontological model which, in the case of ontologies (see Fig. 7.3), comprises concepts, attributes and object properties shown through the relationships between concepts. The formalized conceptual model of the U Ontology is shown in Fig. 7.10. To avoid cluttering the conceptual model diagram, the class diagram shows only the concepts in Fig. 7.10 which, in normal conversion, comes with attributes. The concepts, attributes and their relationships with other concepts are presented in this section to provide an overview of the U Ontology.

U Ontology vocabulary can be classified into the following groups, based on their objectives and functionality: "*Concepts to represent Analysis Metadata*", "*Core Concepts to represent Ontology Usage Analysis*", and "*Concepts to represent Knowledge Patterns*". The constituent concepts, attributes and relationships with other concepts for each group are presented in the following subsection.

It is important to note that the word "concept" will be used here in two contexts; first, to represent the concept defined in the U Ontology and second, to refer to the concepts of the domain ontology being analysed. To avoid homonymity, "Concept/concept" will be used to refer the U Ontology concepts and CONCEPT will refer to the concepts of the domain ontology being analysed. Similarly RELATIONSHIP and ATTRIBUTE refer to the object properties and data properties of the domain ontology, and where required, TERM is used to refer to ontology components.

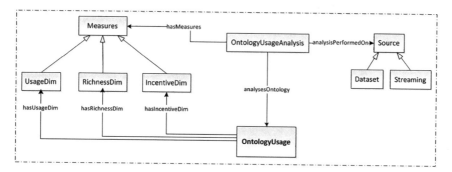

Fig. 7.11 Concepts describing analysis metadata

7.6.1.1 Concepts to Represent Analysis Metadata

This group of concepts represents the portion of the U Ontology conceptual model which deals with concepts pertaining to the analysis of metadata, which are high level concepts of the U Ontology and central to `OntologyUsageAnalysis`. These concepts, which represent the ontology usage analysis domain, create three relationships with `Measure`, `Source` and `OntologyUsage` concepts through `hasMeasure`, `analysisPerformedOn` and `analysesOntology` respectively, as shown in Fig. 7.11. As mentioned in the previous section, usage analysis is measured from three dimensions, therefore `Measure` has three sub-concepts: `UsageDim`, `RichnessDim` and `IncentiveDim` which are linked with `OntologyUsage` through `hasUsageDim`, `hasRichnessDim` and `has IncentiveDim` relationships, respectively. Dataset and Streaming are the sub-concepts of Source for identifying the data source being used to perform the usage analysis.

7.6.1.2 Core Concepts to Represent Ontology Usage Analysis

This group of concepts represents the core concepts of the U Ontology which cover the usage analysis portion of ontology usage analysis. These concepts are divided into three areas, each represented by a dotted rectangular box differentiated by colour, as shown in Fig. 7.12. Each is described as follows:

- The left dotted area (in light green) covers the concept that analysis uses different CONCEPTS. As mentioned previously, CONCEPTS are analysed from different aspects (i.e. CUT), therefore a number of concepts are linked with `ConceptUsage`. `ConceptUsage` represents the CONCEPT and is linked with `Vocab`, `Relationship`, `Attribute` through `hasVocab`, `hasRelation`, and `hasAttribute`, respectively. `ConceptUsage` is linked with `Concept Richness` to specify the richness value which quantifies the structural characteristic of CONCEPT. Likewise, `RelationshipValue` and `AttributeValue`

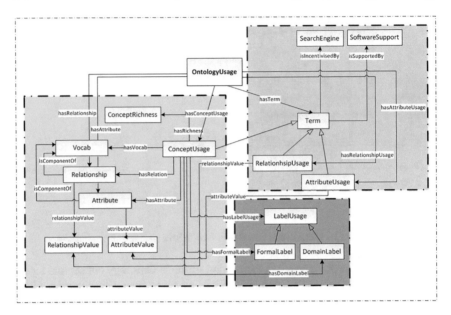

Fig. 7.12 Concepts describing ontology components (concept, relationship, and attribute) analysis

quantify the richness value of object properties and attributes of the domain ontology, respectively.

- The right bottom dotted area (in pink) covers the use of labelling properties for CONCEPT. DomainLabel and FormalLabel are the sub-concepts of LabelUsage and capture the use of the domain ontology defined and the label properties of the W3C-based vocabularies, respectively. ConceptUsage is linked with FormalLabel and DomainLabel using hasFormalLabel and hasDomainLabel properties to specify the use of different labelling properties for CONCEPT.
- The right top dotted area (in light blue) covers concepts related to the usage measurement of RELATIONSHIPS and ATTRIBUTES of the domain ontology. It also includes concepts related to Incentive measurements for the TERMs of the domain ontologies. The Term concept represents TERMs and subsumes ConceptUsage, RelationshipUsage and AttributeUsage. Further Term is linked with SearchEngine and SoftwareSupport through isIncentivisedBy and isSupportedBy properties, respectively to specify which means are used to measure the incentives (commercial benefits).

7.6.1.3 Concepts to Represent Knowledge Patterns

This group of concepts represents the portion of the U Ontology conceptual model which deals with the representation of the knowledge patterns in the dataset, as shown

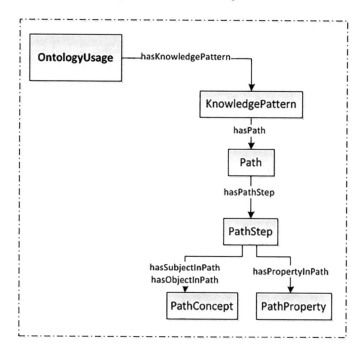

Fig. 7.13 Concepts to represent knowledge patterns in dataset

in Fig. 7.13. As discussed earlier, `KnowledgePattern` represents the prominent usage patterns that invariantly prevail in the dataset.

`KnowledgePatterns` has several `Path` linked through `hasPath` to specify the unique instance of the usage pattern in the `Source`. A `Path` is comprised of several `PathSteps` which essentially represents a triple. In order to capture and represent the subject and object of `PathStep` (triple), `hasSubjectInPath` and `hasObjectInPath` links with `PathConcept` to specify the CONCEPTs. The predicate of `PathStep` is specified by `PathProperty` linked through `hasPropertyInPath`.

The second set of activities in the formalization phase relate to Integration, which is discussed in the next subsection.

7.6.2 Integration with Other Ontologies

The two identified ontologies which conceptualize the domain relevant to OUA, as discussed in Sect. 7.5.2, are the Ontology Metadata Vocabulary (OMV) and the Ontology Application Framework (OAF). OMV enables the specification of the metadata of ontology which includes the ontology conceptualization model (`OntologyBase`), ontology documentation (`OntologyDocument`), tools,

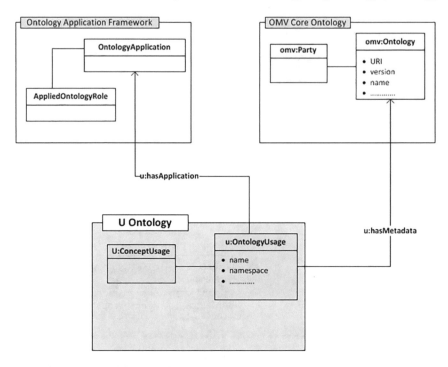

Fig. 7.14 Integration of the U ontology with other ontologies

language, methodology and organization involved in developing and maintaining ontology, whereas the OAF specifies the application areas in which ontologies are used and the roles played by ontologies.

The U Ontology provides interlinking properties, as shown in Fig. 7.14, to integrate these two ontologies with the U Ontology to allow the reuse of the concepts defined in them. The details are as follows:

- The metadata of the domain ontology being analysed is provided by linking the U Ontology's `OntologyUsage` concepts with `omv:OntologyDocument` through `hasMetadata` property. Instead of reinventing the concepts needed to describe the ontology metadata, the OMV ontology's concepts are used for that purpose.
- The Ontology Application Framework (OAF) is used for specifying the application areas in which the domain ontology that is being analysed is used. The U Ontology provides a property `hasApplication` to integrate the U Ontology's `OntologyUsage` with the `OntologyApplication` concept of OAF.

In addition to these two ontologies which are interlinked with the U Ontology, a few terms (URIs) from other common vocabularies/ontologies are used in the U Ontology, as described in Table 7.2.

Table 7.2 Reused terms

Vocabulary	Dublin core
Namespace	http://purl.org/dc/elements/1.1/
Term	title
URI	http://purl.org/dc/elements/1.1/title
Label	Title
Definition	A name given to the resource. Typically, a title will be a name by which the resource is formally known
Vocabulary	Dublin core
Namespace	http://purl.org/dc/elements/1.1/
Term	description
URI	http://purl.org/dc/elements/1.1/description
Label	Description
Definition	An account of the resource. The description may include but is not limited to: an abstract, a table of contents, a graphical representation, or a free-text account of the resource
Vocabulary	RDFS
Namespace	http://www.w3.org/2000/01/rdf-schema#
Term	label
URI	http://www.w3.org/2000/01/rdf-schema#label
Label	Label
Definition	Used to provide a human-readable version of a resource's name
Vocabulary	RDFS
Namespace	http://www.w3.org/2000/01/rdf-schema#
Term	comment
URI	http://www.w3.org/2000/01/rdf-schema#comment
Label	Comment
Definition	Used to provide a human-readable description of a resource
Vocabulary	FOAF
Namespace	http://xmlns.com/foaf/0.1/
Term	name
URI	http://xmlns.com/foaf/0.1/name
Label	Name
Definition	A name for something that is written in a simple textual string

7.7 Implementation Phase: Ontology Implementation

The U Ontology is intended for use on the Web (based on Semantic Web Architecture) to enable users to access usage-related information about ontologies. Therefore, it is desirable that the developed ontology should be able to make use of existing ontology tools such as OWL Reasoner and Triple store, and should be based on the formalism that is largely supported by the community. Based on the literature review on the formalism used for ontologies of a similar nature (such as GoodRelations [142], DQM [103]), OWL DL expressivity is used for the U Ontology. Therefore, for the implementation of the U Ontology, OWL-DL syntax is used which comprises the following language elements:

- `owl:Ontology`
- `owl:Class`
- `owl:ObjectProperty`
- `owl:DatatypeProperty`
- `rdfs:subClassOf`
- `rdfs:subPropertyOf`
- `rdf:datatype`
- `rdf:type`
- `rdfs:domain`
- `rdfs:range`

The choice of the above-mentioned language elements will allow users to use the OWL-DL syntax for RDFS elements. This encoding approach limits the ontology coding to RDFS elements, which is the subset of closure of OWL DLP, and RDFS-based reasoners can be used for inferencing [73]. As suggested in [142] this ontology implementation approach will allow *"the ontology to be used with other OWL-DL ontologies and knowledge bases without making the resulting ontology become OWL Full"*. In encoding, the use of `rdfs:domain` and `rdfs:range` are used to facilitate the data creation, generation and population process by developing a user interface and input form. They should not be used to compute the inference closure by the repositories.

Certain decisions are made pertaining to the specification of the datatype properties of the U Ontology concepts. The following datatype properties are used in the definition of each concept, particularly for those concepts that represent the URIs of the domain ontology, CONCEPTS, RELATIONSHIPS, and ATTRIBUTES such as `ConceptUsage`, `Relationships`, and `Term`.

- `name`
- `termURI`
- `description`
- `prefix`
- `label`
- `comments`

`termURI` attribute and other datatype properties which represent the URIs of the domain ontologies are defined with datatype `xsd:anyURI` to allow the specification of CONCEPTs, RELATIONSHIPs and ATTRIBUTE URIs as objects. For the encoding of the U Ontology, Protégé [168] is used as an ontology editor and provides all the necessary services needed for the construction of ontologies. A number of reasoners provide plug-ins for Protégé, which makes it easy for a developer to validate the conceptual model and perform consistency checks to resolve discrepancies which arise during encoding.

7.8 Recapitulation

In this chapter, the Ontology Usage Ontology (U Ontology) for the *Representation* phase of the OUSAF was presented. A customized development methodology for the development of the U Ontology was adopted, based on three existing methodologies, comprising *Specification, Conceptualization, Formalization* and *Implementation* phases. In the specification phase, motivational scenarios and competency questions are developed to define the scope of the ontology and elicit the requirements of different stakeholders. Based on the identified requirements, the domain knowledge controlled vocabulary is developed as part of the conceptualization framework. Ontologies that are relevant to ontology usage and which overlap with its domain knowledge are identified to be considered for reuse. Based on the terms defined in the controlled vocabulary, a conceptual model is presented to facilitate the formalization of the model. In the formalization phase, the conceptual model of the U Ontology is formalized using the Unified Modeling Language (UML). The models components such as concepts, relationships and attributes were discussed. Ontologies identified for reusability are integrated with the U Ontology to access their components. In the final implementation phase, the U Ontology is encoded in OWL-DL syntax using Protege ontology development tools.

The next chapter focuses on the utilization phase of the OUSAF, which is the implementation of the U Ontology, to demonstrate the application of Ontology Usage Analysis.

Chapter 8
Utilization Phase: Utilization of OUSAF

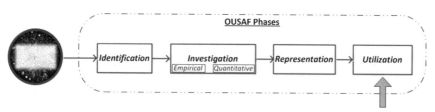

Focus of the Chapter

8.1 Introduction

[1]To demonstrate the utilization of the OUSAF, in this chapter, a methodological approach is adopted which provides a systematic flow of activities and the interaction between different components to analyse the utilization. This methodological approach is presented in Sect. 8.2. Section 8.3 presents details on the construction of the dataset that is used to demonstrate the utilization phase. In Sects. 8.4–8.6, the utilization of the different phases of the OUSAF is presented. Section 8.7 summarizes the achievements of the utilization phase. Section 8.8 concludes the chapter.

---

[1]Parts of this chapter have been:

1. Republished with permission of John Wiley & Sons from Making sense from Big RDF Data: OUSAF for measuring ontology usage, Jamshaid Ashraf, Omar Khadeer Hussain, Farookh Khadeer Hussain, Volume 45, Issue 8, Copyright 2014 John Wiley & Sons, Ltd; permission conveyed through Copyright clearance centre.

2. Reprinted from Knowledge-Based Systems Volume 80, Jamshaid Ashraf, Elizabeth Chang, Omar Khadeer Hussain, Farookh Khadeer Hussain, Ontology usage analysis in the ontology lifecycle: A state-of-the-art review, pp. 34–47, 2017.

© Springer International Publishing AG 2018
J. Ashraf et al., *Measuring and Analysing the Use of Ontologies*, Studies
in Computational Intelligence 767, https://doi.org/10.1007/978-3-319-75681-3_8

8.2   Approach to Demonstrate the Utilization of the OUSAF

To demonstrate the utilization of OUSAF, it is intended to show by performing a certain set of activities, how the required output is obtained for a use case and how this can be used by the specified users to achieve their specified goals (requirements). The term *usability* refers to how the results obtained from the OUSAF can be employed by the users and the term *adequate* indicates whether the results are sufficient to be useful. There are two observations about the underlined words; first, a methodological approach is required so that for each type of user (based on their requirements), the OUSAF can be accessed and results obtained; second, a qualitative discussion will take place to examine whether the results obtained using the framework address the aim of the user. The first observation is implemented by forming a systematic approach to demonstrate the utilization of the framework, and for the second observation, a discussion is presented to explain the usefulness and adequacy of the results. Figure 8.1 shows the infrastructural components, the flow of information, and the applicable observation to demonstrate the utilization of the OUSAF. The flow is as follows: based on the users' requirements (using use cases), the OUSAF is accessed to generate the output accessible through the U Ontology which is then assessed for usability and adequacy.

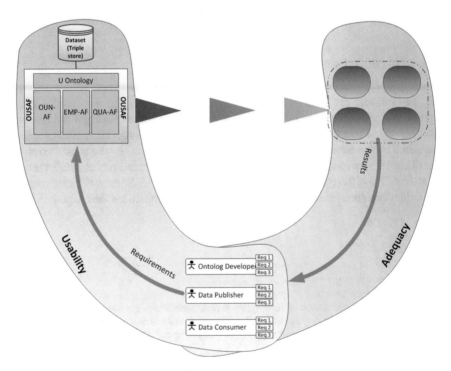

Fig. 8.1 Approach for measuring usability and adequacy

Table 8.1 Components of OUSAF

#	Contribution	Criteria	Description
1	A framework for ontology identification (i.e. OUN-AF)	Analyse the utilization, usability and adequacy of the obtained results	-specify the different types of users of the identification framework - specify the use cases for each user type - analyse the adequacy of the framework in implementing the use cases
2	A framework for empirically analysing the domain ontology usage (i.e. EMP-AF)	Analyse the utilization, usability and adequacy of the obtained results	-specify the different types of users of empirical analysis - specify the use cases for each user type - analyse the adequacy of the framework in implementing the use cases
3	A framework for quantitatively analysing the use of ontologies (i.e. QUA-AF)	Analyse the utilization, usability and adequacy of the obtained results	-specify the different types of users of quantitative analysis - specify the use cases for each user type - analyse the adequacy of the framework in implementing the use cases
4	Formalization of the conceptualized ontology usage analysis domain (U Ontology)	Evaluate the quality of the U Ontology	- specify the methodology which will be used for evaluation - specify the aspects from which ontologies need to be analysed - the methods to analyse the aspects

The components of the OUSAF and the criteria used to analyse the utilization of each component are presented in Table 8.1.

Components 1, 2 and 3 will be analysed in Sects. 8.4, 8.5, and 8.6, respectively. The fourth component, which is the U Ontology, is evaluated in Chap. 9.

8.2.1 Sequence of Activities Involved in Analysing the Utilization of Each Component of the OUSAF

The flow of activities followed in the methodological approach to demonstrate the utilization of each component is shown in Fig. 8.2 and its component description is discussed below.

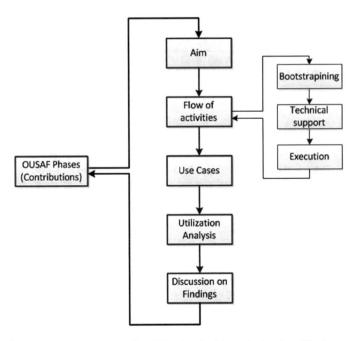

Fig. 8.2 Components and sequence of activities involved in analysing the utilization

Aim: As shown in Fig. 8.2, aim specifies the purpose of the contribution and how it impacts the overall proposed solution. Before proceeding with the utilization analysis steps, the aims of the component and the stakeholders (users) who are interacting with the component need to be specified. The stakeholders include both human users and the machine users, because certain contributions are equally accessible to both types of user.

Flow of activities. Flow of activities specifies the sequence in which the activities need to be performed in order to observe the benefits achievable by the component. This helps to communicate the steps, activities, and their sequence to obtain the desired results. This activity comprises several activities which are described below:

- **Bootstrapping**: Bootstrapping refers to the set of tasks to be performed before the particular contribution of a component is analysed. This involves all the technical and non-technical arrangements required to set up the environment to demonstrate the contributions. Generally, it involves preprocessing the input data into a format that is accessible and usable for the respective framework.
- **Technical support**: Technical support refers to the activities that facilitate the computational arrangement for each component. It provides services at infrastructural level to integrate the different frameworks which implement the respective framework/solution of the component. It involves data manipulation, data structure and access to the dataset.

- **Execution**: Execution refers to carrying out the computational task to implement the contribution of the component. Generally, it involves the execution of the framework to generate output that will then be analysed.

Use Cases: A use case scenario represents the series of actions that need to be carried out to address the specific requirement. Through a use case scenario, user requirements are highlighted and using one of the possible approaches, output which addresses the requirement is obtained.

Utilization Analysis: Utilization analysis refers to the activity in which the contribution of a component in question is analysed against the use case. The output achieved through the contribution and the requirements extracted from the use case are analysed to assess the usefulness and adequacy of the obtained results to the user.

Discussion of Findings: Findings represent the conclusive observations made about the specific contribution. They contribute to discussion about the results obtained to assist in summarizing the utilization analysis of the contribution.

In the next section, the dataset used to analyse the use of ontologies is described.

8.3   Dataset for Demonstrating the Utilization of OUSAF

To conduct the empirical study on the Semantic Web data, and analyse the use of ontologies and a specific (focused) domain, a dataset is built to serve as a representative sample of the semantically annotated structured data currently published on the Web. A hybrid crawler is developed for this purpose which crawls the Web based on the specified parameter and populates the data repository (i.e. triple store). This section describes the design and implementation of the hybrid crawler which collects the snippets of structured data that are embedded in the HTML pages by publishers to provide machine-readable information. Using the hybrid crawler, which has crawled approximately 5.2 million documents (mostly HTML pages), 480 million triples[2] were loaded to the triple store to be used for the analysis of domain ontologies. The specific requirements of the data, crawler, its specifications and the implementation of the crawler are described below. This collected dataset will be used in the following sections of this chapter to demonstrate the utilization of the OUSAF.

8.3.1 Data Requirements

The obvious requirements of the data are that it has to be Semantic Web data described using the RDF data model. The common practice of the community is to publish RDF data using Linked Data principles [140] and recommended best practice [85], and to make them available in the form of a dump file for download. As reported in

---

[2]In fact, triples were converted to quad to add the context of the triple, which is discussed in subsequent sections.

[161], most of the datasets which are included in the LOD cloud make minimal or no use of ontologies, which makes the RDF data available as part of the LOD cloud of limited interest due to the shallow representation of ontologies. The trend in the use of domain ontologies on the Web gained momentum when the incentives (see Sect. 6.3.3 for more detail) were available to data publishers in the form of improved visibility in the search engines and applications were being developed to take advantage of the presence of explicit semantics. To collect a dataset that is primarily annotated using domain ontologies to provide Semantic Web data over the Web, therefore, a hybrid crawler needs to be developed. However, the requirement for the crawler which considers the Semantic data annotated with ontologies has certain unique requirements compared to other crawlers. Generally, crawlers can be grouped into the following categories [146]:

- *Topic-focused crawling*: In this category, hyper links (`<a href="internet URL">`) and anchor text (`<a id="anchor">`) are used to identify pages similar to the topic, based on string matching and link analysis algorithms. Such crawlers are proposed in [3, 55].
- *Ontology-Focused Crawling*: In this type of crawler, ontologies are used to match the concept based on the terms defined in the ontology vocabulary. Examples of such crawlers are proposed in [86, 87].
- *Learning-focused crawling*: In this type of crawler, machine learning techniques are employed to learn about the relevant links and draw similarities between pages. Examples of such crawlers are proposed in [16, 217, 226]
- *Semantic Data-focused Crawling*: In this type of crawler, techniques are used to focus on Semantic Web data that is published on the Web. These crawlers crawl the pages (or documents) that are made available on the Web by describing information using the RDF data model. Examples of such crawlers are proposed in [75, 79, 84, 136].

The first three types of crawler focus on crawling Web pages (or documents) based on their similarity to the topic (subject) of the pages. However, each type has a different approach toward deciding which page to consider and how to route the crawling procedure, e.g., crawl-by-depth or crawl-by-breadth. The second type of crawler (i.e. Ontology-Focused Crawling) makes use of ontologies but does not necessarily operate on Semantic Web data as such. In this approach, ontologies are used to find neighbouring and similar concepts that match the topic by allowing the crawling to expand to similar related concepts based on the ontology conceptual model. The last type of crawler (i.e. Semantic Data-focused Crawling) is focused on crawling the RDF documents which are published on the Web. These types of crawler apply a specific filter to consider only those documents that match the criteria, such as MIME type and Content-type.

As mentioned earlier, not all the Semantic Web (RDF) data published on the Web uses domain ontologies to describe information, and more emphasis is placed on publishing structured data on the Web with or without the use of explicit semantics. Merely considering RDF data, which is normally made available in the form of dump files, does not provide a fair representation of the structured data that is annotated using domain ontologies. Therefore, the requirements of the data to be considered for measuring and analysing the use of domain ontologies are as follows:

- Semantic Web data that is based on the RDF data model
- Semantic Web data that is described using domain ontologies
- Semantic Web data that is published either as an RDF document or embedded within HTML pages

A crawler was developed by extending [160] to crawl RDF data, but the results were not encouraging because most RDF documents did not use domain ontologies, except for the use of W3C-based vocabularies such as RDF, RDFS and a few constructs of OWL. The other two vocabularies which had a reasonable presence in RDF documents were FOAF and DC, both of which are considered to be sufficiently well established to provide a textual description of the resources. Another observation was that there is minimal use of out-bound links to external documents (or resources). In light of these observations, a new crawling strategy was devised to focus on the Web pages that have semantically annotated structured data embedded in them. The detail of this strategy is discussed in the next section.

8.3.2   Data Collection Strategy

A crawler was initially implemented by extending the LDSpider [160] but the collected data was not interesting as only 8.52% of the 1.8 million triples were described using authoritative ontologies, excluding W3C-based vocabularies, based on the statistics. To obtain a dataset that addresses the previously discussed requirements, a new strategy was devised to customize a crawler that is capable of collecting the required Semantic Web data. In our previous study [8], it was observed that the new publishing trend is to add Semantic Web data using the RDFa standard, which allows the addition of RDF snippets within exiting HTML pages. In previous research, it was observed that 90% of Semantic data is published using RDFa when it is embedded within existing Web pages and annotated with ontologies.

Based on our experience and considering the recommendation proposed in [259], a data collection strategy was devised as described below.

- Using Semantic Web search engines such as Swoogle, Watson, Sindice, extract the list of domain names (pay-level-domain (PLD)) that publish data annotated using ontologies to generate seed URIs.
- Instead of focusing on the PLD which provides `Content-type: application/rdf+xml`, also consider `application/xhtml+xml`, to include HTML documents which have AN RDFa snippet embedded in them.
- Exclude web pages in which structured data is embedded using a non-RDFa standard such as microdata[3] and microformats.[4]

---

[3]https://www.w3.org/TR/microdata/; retr. 14/11/2017.
[4]http://microformats.org/; retr. 25/11/2017.

Fig. 8.3 Components involved in data collection

- Exclude URI schemes from the seed URI list and a crawler process which includes *ftp, telnet, maitto* and *file*.
- Exclude digital resources such as *image*, *pdf*, and *cvs* files.

The data collection strategy is formed according to the guidelines above and used for the data collection process, which is discussed in the next subsection.

8.3.3 Data Collection Process

The data collection process specifies the steps and components involved in the collection of data using a hybrid crawler. Figure 8.3 shows the components involved in the overall process of data collection. The role of each component is explained below.

The crawler proceeds based on the seed URIs collected through the Seed RUI Builder which is responsible for preparing the seed URI to initiate the crawling process. From the seed URI, the RDF document (or HTML page) is retrieved to obtain the contents of the URIs. The obtained contents are then parsed to transform them into the required format before the context of the retrieved RDF document is specified in the contextualization phase. The contextualized content is subsequently loaded into the triple store for analysis.

8.3.3.1 Seed URL Builder

The crawler operates on a list of unvisited URLs which is known as a *frontier* (i.e. the list). The list of URLs is initialized with seed URLs which are often collected through another program or manually supplied. The quality of the data retrieved by the crawler depends on the quality of the seed URLs. The role of the Seed URL Builder is to provide a list of URLs to initiate the crawl and specify the frontiers. To obtain high quality URLs to provide the frontier for the first round, semantic search engines are accessed to retrieve the URLs of the websites (data publishers) that publish data using domain ontologies. Two semantic search engines, namely Watson [66] and Sindice [266], and one traditional search engine, Google, are accessed using their APIs to obtain a list of URLs, as shown in Fig. 8.4. To retrieve the RDF document `filetype:rdf` in Google, the attribute of advanced search is used to narrow the search to only documents with the specified extension. To specify the quality, the

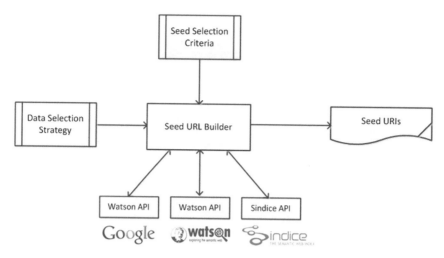

Fig. 8.4 Seed URL builder

number of namespaces defined in the RDF document (HTML pages), aside from W3C-based vocabularies, are measured and a weight $Seed_w$ is specified.

8.3.3.2 Semantic Document Downloader

Following the general flow [218] in each round, the URL is picked from the frontier to retrieve the document corresponding to the URL using the HTTP request. The retrieved document is parsed to obtain the content and the links to the external documents. These links are then evaluated for inclusion in the frontier, based on their quality value.

8.3.3.3 Snippet Extractor

Snippet Extractor extracts the RDFa snippets from the HTML pages and transforms each snippet into an RDF/XM-based RDF document. Using the parser, Snippet Extractor retrieves the content of the document and extracts the triples, annotating the information. Any23[5] and RDFaDistiller[6] services are used to extract and transform the triples into RDF/XML serialized format.

---

[5]https://any23.apache.org/; retr. 09/11/2017.
[6]https://www.w3.org/2012/pyRdfa/; retr. 18/11/2017.

8.3.3.4 Contextualizer

From a data management point of view, the context of the retrieved documents and their provenance details need to be added to the extracted RDF graph. For this purpose, the Named Graph [53] approach is used to convert the triple into quads by adding a new resource that specifies the context of the transformed graph.

8.3.3.5 Loader

Loader loads the quads into the triple store for usage analysis. Provenance details such as the date and time when the data was collected, the source origin detail such as the PLD and the original data format are gathered.

8.3.4 Crawling

Figure 8.5 shows the sequential flow of the crawler implemented to collect Semantic Web data to measure ontology usage. The crawler starts by populating the Seed URLs. The URLs to be visited are called frontiers and in one round, all these frontiers are covered. The crawler retrieves the `robots.txt` file from each URL to ensure the required politeness in the crawling process. Filter criteria (mentioned under data collection strategy) are used to decide how to fetch the RDF documents (HTML pages) from the PLD. If the required page is allowed to be fetched, it is downloaded from the Web to extract its content. Web pages with RDFa snippets embedded in them are crawled and the RDF triples extracted using RDFa parsers. From the extracted RDF/XML graph, URLs (the resources URI) referring to external resources are evaluated and enqueued to the seed URL list. The parsed RDF/XML graph is then contextualized by converting the triple into a quad and provenance details are added before the graph is loaded to the triple store.

A crawling policy was developed to run the crawler without adding unnecessary load on the server from which the Semantic Web data is being crawled. First, the restrictions and permissions outlined in the `robots.txt` file were strictly observed. In cases where these were not published by the server, politeness in crawling as suggested in [259] was incorporated. In fetching multiple pages from the same PLD, three fetch requests per second were used, and the crawler was paused for 10 s after 1000 requests in an attempt to ensure the server was not over busy with the fetching routine. REST-based services were used to transform the extracted RDFa snippets to RDF/XML graph, and only the final RDF graph was loaded to the triple store. While accessing the Any23 and RDFaDistiller REST-based services, a delay of one second was applied to avoid overloading the servers hosting the service. The hybrid crawler was developed based on LDSpider and deployed on an Intel Core 2 PC with 2GB RAM.

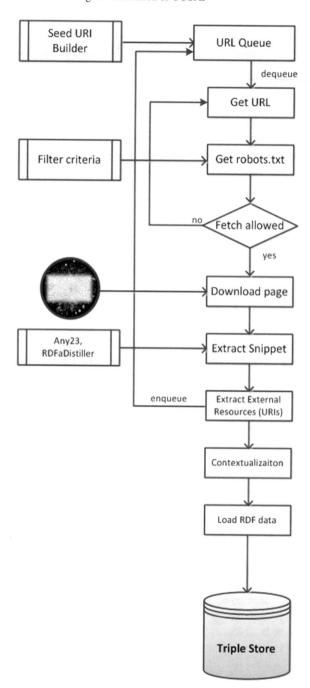

Fig. 8.5 Basic flow of the Crawler's activities

Table 8.2 Content type of HTTP response

Content type	% of documents
text/html; charset=iso-8859-1	7.32
application/octet-stream	0.24
application/rdf+xml	22.93
application/xml	0.73
text/html	22.20
text/html; charset=UTF-8	40.73
text/plain	2.44
text/xml	2.20
text/xml; charset=UTF-8	1.22

8.3.5 Dataset Statistics

To obtain the required dataset to measure ontology usage analysis, a hybrid crawler was developed based on LDSpider and deployed on an Intel Core 2 PC with 2GB RAM, as noted in the previous subsection. The data was collected over a period of one month at various intervals, taking 186 machine hours in total. Using 12,000 seed URLS during the crawl, 480 quads were loaded from 5.2 million documents (90% HTML pages). The first two Linked Data principles (see Sect. 1.2 for more detail) require the use of HTTP URI to uniquely name resources and make them accessible through HTTP request as defined by RFC3986.[7] In URI lookup, 79.27% of URIs returned the 200 OK code which is the standard response for successful HTTP requests. This may seem surprising, because datasets in the Semantic Web are normally available through redirect, but in the case of the e-Commerce web of data, most of the structured information is published using RDFa, embedded within HTML documents, thus making it available through standard HTTP request. The 5XX code represents a server error and 1.22% of the URIs returned the 5XX code. 18.05% of URIs returned the 404 Not Found, code which indicates that the requested resource could not be found but might be available in the future. 1.46% of URIs resulted in a redirect with the 302 code to reroute the request to a non-information resource at an alternative URI. Table 8.2 lists the percentage of content-type found in the documents. As mentioned earlier, most of the data sources use RDFa, therefore documents are considered to be both information and non-information sources with application/rdf+xml content type.

Using the approach presented in Sect. 8.2, the utilization of the OUN-AF, EMP-AF, and QUA-AF is presented in the following sections.

---

[7]https://tools.ietf.org/html/rfc3986; retr.; 19/11/2017.

8.4   Utilization of the Identification Framework (OUN-AF)

8.4.1   Aim

The aim of the identification phase is to identify the use of different ontologies and the interlinking between them based on their instantiation; in other words, which ontologies are being used to semantically describe domain specific entities at the instance level. It also identifies usage patterns that are prevalent across different data publishers and similarly, the co-usage patterns of different ontologies. The OUN-AF is used to obtain these insights.

8.4.2   Flow of Activities

The OUN-AF is based on the affiliation network which comprises two sets of nodes: ontologies and data source. OUN is constructed based on the data sources and ontologies present in the dataset. OUN is used to measure the use of ontologies by different data sources and also to identify the co-affiliation between different ontologies based on their co-usage in describing the information of a data source. The projection approach is used to transform the two-mode network into a one-mode network and measure the co-affiliation factor.

8.4.3   Use Cases

The three types of user are defined in Sect. 1.7; based on their function and role, they each require a different set of ontology identification information. The information requirements of each type of user are described in the identification phase and then used to demonstrate the utilization of the framework.

Ontology developer's requirements

Req. (1) What is the level of usage of a given ontology?

Req. (2) Is the given ontology being used alone or with other ontologies, and if the latter, what are they?

Data consumer's requirements

Req. (3) Which ontologies are being used in a given domain?

Req. (4) Which data sources are using a given ontology to publish their information?

Data publishers' requirements

Req. (5) Which cohesive groups of ontologies have similar usage?

The requirements listed above are used to analyse the utilization of the OUN-AF.

```
1   PREFIX uo: <http://example.uontology.org/v1#>
2
3   SELECT ?ontology ?oname ?ouri ?oprefix ?ousers
4   WHERE {
5           ?ontology rdf:type uo:OntologyUsage;
6                     uo:name    ?oname;
7                     uo:uri     ?ouri;
8                     uo:prefix  ?oprefix;
9                     uo:hasUsers ?ousers.
10          FILTER regex(?oprefix, "foaf", "i")
11  }
```

Fig. 8.6 SPARQL query to display the list of ontologies and their usage

8.4.4 Utilization Analysis

8.4.4.1 Req. 1: What is the Level of Usage of a Given Ontology?

In this requirement, the user is interested in knowing how many data sources are using a particular ontology. This requires the return of a number of data sources (ontology users) which have used the ontology components (at least one term of the ontology) to describe the information published on the Web.

The Ontology Usage Distribution (OUD) metric (Sect. 4.7.1) of the OUN-AF measures the number of different data sources a particular ontology has. This measure is also represented in the U Ontology (Chap. 7) which conceptualizes the domain of ontology usage analysis. The OntologyUsage concept has the hasUsers attribute which captures the value of the OUD metric. Figure 8.6 displays the SPARQL query which instructs the U ontology to retrieve the usage of the given domain. The filter clause of the query represents the input of the user, indicating the particular usage in which they are interested. In the query, the name, URI, prefix and number of data sources using the given ontology (e.g. FOAF ontology) are displayed.

From Fig. 8.6, it can be seen that the U Ontology represents and captures concepts and attributes from which the user can obtain information about the use of different ontologies in the dataset.

Figure 8.7 displays the results of the query in Fig. 8.6. In order to display other ontologies apart from *foaf*, the FILTER clause was removed when the query was executed. The results display the object reference, name, namespace URI, prefix and number of users (data publishers who have used the ontology) for each ontology. As can be seen, there are 134 data publishers in the data set who have used the Open Graph Protocol vocabulary (5th row).

This insight regarding the usage of different vocabularies enables users (particularly ontology developers) to learn the present adoption level and uptake of ontologies on the Web. Developers can learn from the ontologies which have a good adoption rate by studying their structural and semantic characteristics and applying them to

| ← → | 🔵 ▾ | http://localhost:8080/oua/analysisquery/req1.jsp | → | 🔍 ▾ Google | 🔍 🏠 🔳 ▾ |

Ontology Usage Analysis Framework (OUSAF)

U Ontology

ontology	oname	ouri	oprefix	ousers
http://data.uontology.org/oua/ont/gr004	GoodRelations	http://purl.org/goodrelations/v1#	gr	198
http://data.uontology.org/oua/ont/fo002	Friend of a Friend	http://xmlns.com/foaf/0.1/	foaf	145
http://data.uontology.org/oua/ont/vc012	vCard	http://www.w3.org/2006/vcard/ns#	vcard	140
http://data.uontology.org/oua/ont/xv014	vocabulary (google)	http://rdf.data-vocabulary.org	v	137
http://data.uontology.org/oua/ont/og005	open graph protocol (facebook)	http://opengraphprotocol.org/schema/	og	134
http://data.uontology.org/oua/ont/dc003	Dublin Core	http://purl.org/dc/terms/	dc	133
http://data.uontology.org/oua/ont/re010	review	http://purl.org/stuff/rev#	rev	105
http://data.uontology.org/oua/ont/ya021	yahoo	http://search.yahoo.com/searchmonkey/commerce/	yahoo	97
http://data.uontology.org/oua/ont/...	product ontology	http://www.productontology.org/id/	pto	78

Fig. 8.7 Result of SPARQL query shown in Fig. 8.6

their own ontology development process. Data consumers are provided with a list of well-adopted ontologies to consider for their own semantic annotation needs.

8.4.4.2 Req. 2: Is the Given Ontology Being Used Alone or with Other Ontologies, and if the Latter, What Are They?

In this requirement, the user would like to know the ontologies that are being co-used with the given ontology. This enables the user to identify the ontologies that cover concepts related to their domain and are frequently used by the community of Semantic Web data publishers.

The Ontology Usage Network (OUN) is a two-mode network with relationships (edges) between distinct types of node. In the case of Req. 2, it is necessary to know the relationships between the same types of node to understand which ontologies are being co-used with a given ontology. Using the projection technique (Sect. 4.5.2), an ontology-to-ontology network is obtained which represents the relationships between ontologies. As explained in Sect. 4.8.4, two ontologies in the projected one-mode network are linked only if both have been used by the same data source, which shows their co-usability factor in the network.

```
1   PREFIX uo: <http://example.uontology.org/v1#>
2
3   SELECT ?oname
4   WHERE {
5           <http://data.uontology.org/oua/ont/gr004>   rdf:type uo:OntologyUsage;
6                       uo:isCoused      ?coonto.
7           ?coonto   uo:name          ?oname.
8   }
```

Fig. 8.8 SPARQL query to display the names of the ontologies being co-used

The U Ontology provides the isCoused relationship which links two ontologies if they have an edge in the ontology-to-ontology one-mode (projected) network.

Figure 8.8 shows the SPARQL query which lists the names of the ontologies that are being co-used with a given ontology. In the listing, <http://data.uontology.org/oua/ont/gr004> is the URI of the given ontology (e.g. FOAF). Using uo:isCoused object property, the URIs of the ontologies which are being co-used are obtained and uo:name data property displays the name of the ontologies.

The query listed in Fig. 8.8 shows that the U Ontology model is able to capture information regarding the co-usability factor obtained by projecting OUN to an ontology-to-ontology network.

The query shown in Fig. 8.8 requires all other ontologies which have been used along with a particular ontology, i.e. "gr", to be displayed. Figure 8.9 shows the names of the ontologies which have been used by different data publishers to semantically describe e-Commerce-related information. Knowing which other ontologies are being used with a given ontology helps ontology developers to know what other ontologies are sharing the conceptual description related to the domain being captured by the given ontology. It provides data publishers with a list of ontologies they should consider when deciding on ontologies for their information annotation needs.

8.4.4.3   Req. 3: Which Ontologies Are Being Used in a Given Domain?

In this requirement, the user is interested to know the different ontologies that are currently being used on the Web in a specific application area. This is a common requirement for all types of users because it provides high level but useful information about ontology usage. This requirement is closely matched with Req. 1 (Sect. 8.4.4.1); however, here the user is interested to know all the ontologies that are being used in the dataset. Figure 8.10 lists the query which retrieves all the ontologies that are being used in the dataset. The name, prefix, URI, and usage of all ontologies are obtained to provide the required information for users.

This query is similar to the one shown in Fig. 8.6, and the result obtained is similar to that shown in Fig. 8.7.

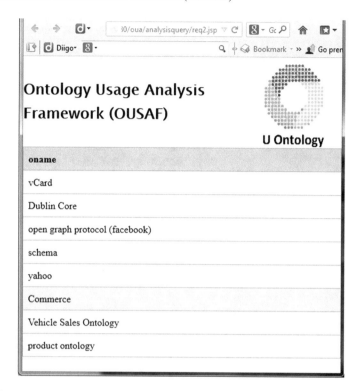

Fig. 8.9 Result of SPARQL query shown in Fig. 8.8

```
1   PREFIX uo: <http://example.uontology.org/v1#>
2
3   SELECT ?ontology ?oname ?ouri ?oprefix ?ousers
4   WHERE {
5           ?ontology rdf:type uo:OntologyUsage;
6                     uo:name     ?oname;
7                     uo:uri      ?ouri;
8                     uo:prefix   ?oprefix;
9                     uo:hasUsers ?ousers.
10  }
```

Fig. 8.10 SPARQL query to display the name of the ontologies present in the dataset

```
1    PREFIX uo: <http://example.uontology.org/v1#>
2
3    SELECT ?oname ?dsname ?dsind
4    WHERE {
5            ?ontology rdf:type uo:OntologyUsage;
6                          uo:prefix     ?oprefix;
7                          uo:name       ?oname;
8                          uo:isUsedBy   ?ds.
9            ?ds           uo:name       ?dsname;
10                         uo:industry   ?dsind.
11           FILTER regex(?oprefix, "foaf", "i")
12   }
```

Fig. 8.11   SPARQL query to display the name of data sources which have used a given ontology

8.4.4.4   Req. 4: What Data Sources Are Using a Given Ontology to Publish Their Information?

In this requirement, the user is interested in knowing about the different data sources (pay level domains) that are using the given ontology. The second set of OUN nodes represent the data sources which have used the ontologies to describe information. In the U Ontology, the information about the different data sources is represented through the `DataSource` concept. The `OntologyUsage` concept is linked to the `DataSource` concept through `isUsedBy` relationships which allow the data sources using the given ontology to be specified.

Figure 8.11 displays the SPARQL query for accessing the U Ontology to obtain the list of data sources which are using the particular ontology and displays the name of the ontology, the data source name (which is actually the URL), and the industry of the data source. This information helps the data consumer to know more about the adoption and uptake of the ontology in real world implementation.

Figure 8.12 displays the list of data sources and the industry to which they belong. The query has returned the names of data sources that have used *foaf* (Friend of a Friend) and their respective industry (application domain). Knowing who is using a particular ontology and their domain helps ontology developers to perform a detailed analysis of these data sources to investigate exactly how the ontology components are being used.

8.4.4.5   Req. 5: What Cohesive Groups of Ontologies Have Similar Usage?

In this requirement, the user would like to know the various cohesive groups in the dataset based on their co-usage. This identifies the ontologies which have some commonality (which could be semantic similarity in terms of describing related but different concepts) to enable users to understand or analyse their characteristics. In

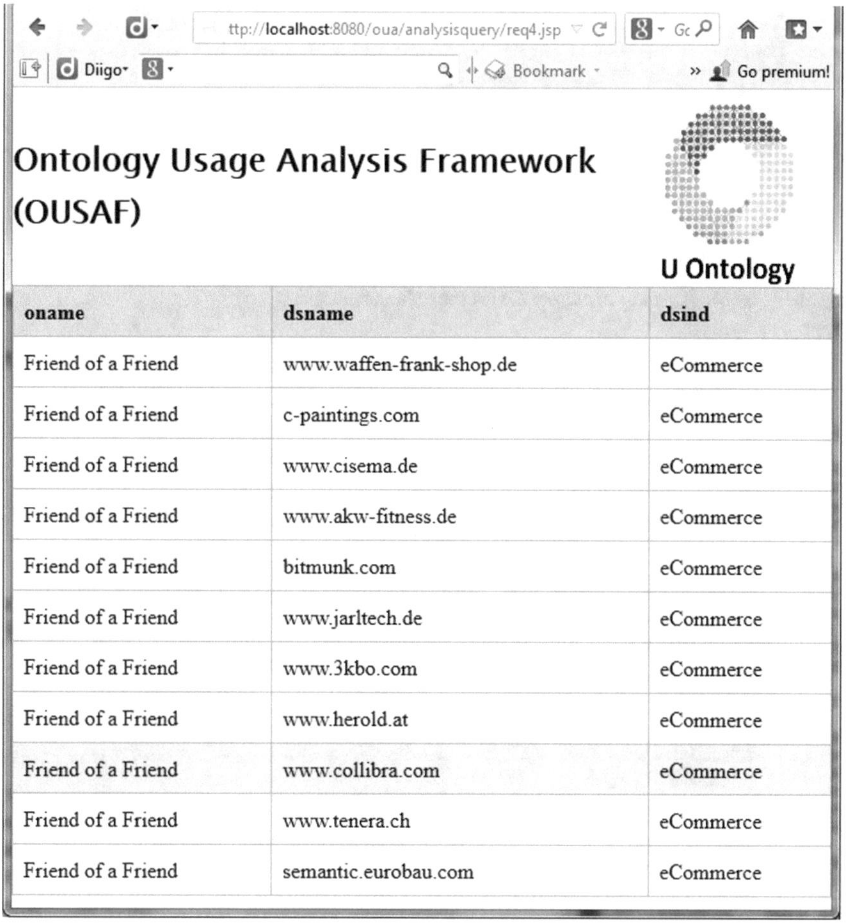

Ontology Usage Analysis Framework (OUSAF)

oname	dsname	dsind
Friend of a Friend	www.waffen-frank-shop.de	eCommerce
Friend of a Friend	c-paintings.com	eCommerce
Friend of a Friend	www.cisema.de	eCommerce
Friend of a Friend	www.akw-fitness.de	eCommerce
Friend of a Friend	bitmunk.com	eCommerce
Friend of a Friend	www.jarltech.de	eCommerce
Friend of a Friend	www.3kbo.com	eCommerce
Friend of a Friend	www.herold.at	eCommerce
Friend of a Friend	www.collibra.com	eCommerce
Friend of a Friend	www.tenera.ch	eCommerce
Friend of a Friend	semantic.eurobau.com	eCommerce

Fig. 8.12 Result of SPARQL query shown in Fig. 8.11

the OUN-AF, the Cohesive Subgroups metric is defined to measure the k-core of the Ontology Co-Usage network. The U Ontology provides the attribute to specify the k-core value to which the ontology belongs. The `OntologyUsage` concept has the attribute `hasCousedValue` to represent the k-core value to which it belongs. The value is the degree of the node (given ontology) of the projected Ontology Co-Usage network which indicates how many ontologies it is being co-used with.

Figure 8.13 lists the SPARQL query which displays the k-core value of the ontologies. The ontologies with the same k-core value belong to the same cohesive group. To obtain the ontologies belonging to a particular cohesive group, the *?kcore* variable can be used to limit the ontologies belonging to a group.

Figure 8.14 displays the names of the ontologies and the cohesive group to which they belong. It can be seen from the list which ontologies have a similar usage in the

```
1   PREFIX uo: <http://example.uontology.org/v1#>
2
3   SELECT ?oname ?kcore
4   WHERE {
5           ?ontology rdf:type uo:OntologyUsage;
6                        uo:name         ?oname;
7                        uo:hasCousedValue  ?kcore;
8   }
```

Fig. 8.13 SPARQL query to extract the k-core value ontologies

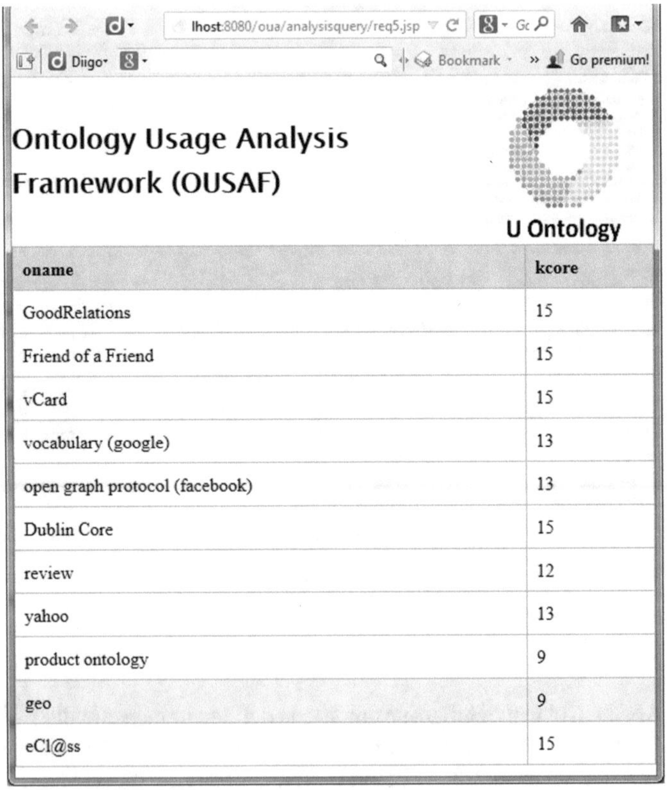

oname	kcore
GoodRelations	15
Friend of a Friend	15
vCard	15
vocabulary (google)	13
open graph protocol (facebook)	13
Dublin Core	15
review	12
yahoo	13
product ontology	9
geo	9
eCl@ss	15

Fig. 8.14 Result of SPARQL query shown in Fig. 8.13

dataset. This insight into the presence of cohesive groups helps ontology developers to know the co-usability factor among the different ontologies.

8.4.5   Discussion of Findings

In this section, the usability and adequacy of the results from the identification phase are presented. Use cases which represent the frequently occurring requirements of the users are presented to analyse the usefulness and adequacy of the results obtained through OUN-AF. For each requirement, the ontology identification phase, its computational model and applicable metrics are discussed to describe their capability to address these requirements. For users to make use of the ontology usage analysis, the U Ontology is accessed to obtain the required information by posing the SPARQL queries for each requirement. Based on the described use cases and the solution offered by the OUN-AF and U Ontology, the findings are summarized in the following points:

1. The OUN-AF is able to provide the method and techniques necessary to address the requirements of the identification phase of the OUSAF. The **OUN-AF, its techniques and methods are therefore capable of providing the required insight into ontology identification**.
2. The U Ontology is able to represent and capture the concepts pertaining to the ontology identification phase. **The information required for each use case was retrieved by accessing the U Ontology, therefore the conceptual model formalizing the domain of ontology usage is considered to be effective in addressing users requirements**.

8.5   Utilization of the Empirical Analysis Framework (EMP-AF)

8.5.1   Aim

The aim of the investigation phase of the OUSAF is to analyse the use of ontologies on the Web. To do this, the analysis is performed at two levels: the empirical level and the quantitative level. The EMP-AF empirically analyses ontology usage from different aspects to provide insight into the use of an ontology and its components, including the use of multiple concepts, other ontologies/vocabularies used to describe the entities, and the use of different relationships and data properties to provide factual statements about entities.

8.5.2   Flow of Activities

The ontology identified by the OUN-AF or provided by the user is empirically analysed using the EMP-AF. The framework makes use of the dataset collected by the hybrid crawler presented in Sect. 8.3. The EMP-AF for each concept applies the Concept Usage Template (CUT) which analyses the concepts from many aspects, and metrics are used to measure them. The analysis and ontology components obtained are then populated into the U Ontology for the dissemination of usage analysis.

8.5.3   Use Cases

The EMP-AF empirically analyses the use of ontologies and provides detailed insight into the use of ontologies and their components by different data sources (data publishers). The aspects which are analysed pertaining to concepts are their instantiation, the use of other ontologies to describe the entities instantiated by the concept, the relationship the concept has with other entities, the use of label properties, and prevalent knowledge patterns in the dataset.

The three types of users defined in Sect. 1.7, based on their function and role, each require a different set of information for the ontology usage analysis. The information requirements of each type of user are described below and will be used to demonstrate the utilization of the framework.

Ontology developer's requirements
   Req. (1) What is the adoption level of a given ontology?
   Req. (2) How are the entities of a given concept described?

Data consumers requirements
   Req. (3) How are entities textually described?

Data publishers requirements
   Req. (4) What knowledge patterns are available in the dataset?
   The requirements listed above are used to demonstrate the utilization of the EMP-AF.

8.5.4   Utilization Analysis

8.5.4.1   Req. 1: What is the Adoption Level of a Given Ontology?

In this requirement, the user is interested in knowing how a particular ontology is being adopted by the end user. It could be that the ontology developer is interested in their own developed ontology, or would like to observe the usage trends in similar ontologies. The adoption of an ontology is an generic observation which can include several components to provide a comprehensive overview of how an ontology and its components are being used. Here, the user is interested in knowing the terminologies of the ontology that have some usage on the Web (in real world implementation).

The EMP-AF empirically analyses the use of the various terms of the ontology in the dataset. The CUT template and other metrics defined as part of the EMP-AF generate the terminological knowledge of the ontology that is being used and

```
1   PREFIX uo: <http://example.uontology.org/v1#>
2
3   SELECT ?ocon, ?oatt, ?orel
4   WHERE {
5           ?ontology  uo:prefix      "gr".
6
7           OPTIONAL {?ontology  uo:hasConceptUsage      ?ocon.}
8           OPTIONAL {?ontology  uo:hasAttributeUsahe    ?oatt.}
9           OPTIONAL {?ontology  uo:hasRelationshipUsage ?orel.}
10  }
11  LIMIT 50
```

Fig. 8.15 SPARQL query to display the terms of the ontology which have usage in the dataset

```
1   SQL>SPARQL
2   DEFINE input:inference "http://example.uontology.org/inference/rdf9"
3   SELECT ?tname
4   WHERE{
5       ?ontology  uo:prefix   "gr".
6       ?ontology  uo:hasTerm   ?oterm.
7       ?oterm     uo:name      ?tname.
8   };
```

Fig. 8.16 Query exploiting RDFS entailment rule (rdfs9)

adopted by data publishers. The U Ontology captures the components of the ontology, along with their usage, to provide an overview of their usage uptake. Figure 8.15 displays the SPARQL query for retrieving the terminologies of the ontology which are being used on the Web. Since the objective is to obtain the list of terms which have been used, irrespective of their usage level, the query does not retrieve their usage frequency.

The query shown in Fig. 8.15 lists all the ontology components, including the concepts (classes), object properties (relationships) and data type properties (attributes) of the ontologies which have instantiation in the dataset. The U Ontology model captures all these components through `ConceptUsage`, `RelationshipUsage` and `AttributeUsage` concepts. However, as can be seen in the U Ontology model (Fig. 7.10), these three classes are subclasses of the class `Term`.

This subsumption relationship allows the retrieval of all the instances of the subclasses through the use of RDFS entailment rules. Applying the axiomatic triples available in the ontology and RDFS9 rule set, the implied information at the instance level can be retrieved. The Virtuoso (open source) triple store [212] which provides RDFS rule-based reasoning support, inference context i.e `http://example.uontology.org/inference/rdf9`), is defined to retrieve all the terms of the ontology that have usage through the term concept. Figure 8.16 displays the SPARQL query which retrieves a similar result but through inference.

Figure 8.17 displays the names of the "gr" terms that have been used in the dataset. The result screen is edited to show the concepts, object properties and the attributes in the first, second and third columns, respectively, and provides a comprehensive

U Ontology

ocon	oatt	orel
gr:Offering	gr:validFrom	gr:offers
gr:ProductOrServiceModel	gr:validThrough	gr:hasPOS
gr:BusinessEntity	gr:eligibleRegions	gr:availableAtOrFrom
gr:UnitPriceSpecification	gr:hasStockKeepingUnit	gr:hasBusinessFunction
gr:ProductOrServicesSomeInstancesPlaceholder	gr:availabilityStarts	gr:eligibleCustomerTypes
gr:LocationOfSalesOrServiceProvisioning	gr:hasEAN_UCC-13	gr:acceptedPaymentMethods
gr:QualitativeValue	gr:description	gr:availableDeliveryMethods
gr:QuantitativeValueInteger	gr:name	gr:hasOpeningHoursDayOfWeek
gr:QuantitativeValue	gr:condition	gr:hasOpeningHoursSpecification
gr:TypeAndQuantityNode	gr:hasMPN	gr:typeOfGood
gr:ProductOrService	gr:BusinessEntity	gr:includesObject
gr:QuantitativeValueFloat	gr:hasCurrency	gr:hasPriceSpecification
gr:OpeningHoursSpecification	gr:hasValue	gr:hasManufacturer
gr:DeliveryMethod	gr:hasValueFloat	gr:hasMakeAndModel
gr:BusinessEntityType	gr:hasValueInteger	gr:hasWarrantyScope
gr:PaymentMethodCreditCard		gr:hasInventoryLevel
gr:BusinessFunction		gr:hasWarrantyPromise
gr:WarrantyPromise		gr:deliveryLeadTime
		gr:includes
		gr:weight

Fig. 8.17 Result of SPARQL query shown in Fig. 8.15

list of ontology terms. This insight is useful for all types of users because it provides a consolidated view of the ontology usage. Ontology developers can use it to know which concepts are being instantiated and analyse those which are not being used. This is also useful for implementing changes to the ontology, as it is based on what is being used and ontology developers can choose a suitable approach.

8.5.4.2   Req. 2: How Are the Entities of a Given Concept Described?

In this requirement, the user is interested in knowing how the entities of a specific type (instance of a concept) are being semantically described on the Web. This

```
1    PREFIX uo:  <http://example.uontology.org/v1#>
2
3    SELECT ?rname,  ?rvname,  ?aname,  ?avname
4    WHERE {
5            ?concept  uo:name        "BusinessEntity";
6                      uo:prefix      "gr".
7            {
8            ?concept  uo:hasRelation   ?rrel;
9                      uo:name          ?rname;
10                     uo:isComponentOf ?rvocab.
11           ?rvocab   uo:name          ?rvname.
12           }
13           UNION
14           {
15           ?concept  uo:hasAttribute  ?aatt;
16                     uo:name          ?aname;
17                     uo:isComponentOf ?avocab.
18           ?avocab   uo:name          ?avname.
19           }
20   }
21   LIMIT 50
```

Fig. 8.18   SPARQL query to display the use of different relationships and attributes

information not only assists ontology owners to know the prevalent entity schema, but is also of interest to data publishers and consumers. For ontology developers, it provides insight into the use of different relationships to describe the related entities and their aspects, and attributes to provide factual statements about the entity.

The Concept Usage Template (CUT) of the EMP-AF provides the model to capture the relevant aspects of the entities. The U Ontology represents the components of the CUT to enable users to access those that are relevant. Figure 8.18 lists the SPARQL query which accesses the U Ontology to retrieve the semantic description used to define the entity of a specific type (concept). The query displays the semantic description of the entity which is the instance of the `BusinessEntity` concept of the GoodRelations ontology. In the query, all the object properties and data type properties which have been used in the dataset to describe the entity are displayed. Since it is common practice to reuse terms defined in other ontologies, their respective ontology is also retrieved in the query in addition to term names.

Figure 8.19 displays the properties being used to describe `BusinessEntity` entity. It can be seen that terms from other ontologies are being reused to describe the instance of the entity. This helps ontology developers to know which other concepts are being frequently used to describe that entity, and enables data publishers to know which other terms should be considered for semantic annotation.

rname	rvname	aname	avname
gr:offers	GoodRelations	gr:legalName	GoodRelations
gr:hasPOS	GoodRelations	gr:hasISICv4	GoodRelations
vCard:add	vCard	gr:hasNAICS	GoodRelations

Fig. 8.19 Result of SPARQL query shown in Fig. 8.18

```
1   PREFIX uo: <http://example.uontology.org/v1#>
2
3   SELECT ?flname, ?dlname
4   WHERE {
5           ?concept uo:name      "BusinessEntity";
6                    uo:prefix    "gr".
7           {
8           ?concept uo:hasFormalLabel ?flab;
9                    uo:name           ?flname.
10          }
11          UNION
12          {
13          ?concept uo:hasDomainLabel ?dlab;
14                   uo:name           ?dlname.
15          }
16  }
```

Fig. 8.20 SPARQL query to access the formal labels and domain labels used for a concept

8.5.4.3   Req. 3: How are Entities Textually Described?

In this requirement, the user, particularly the data consumer (application developer), is interested in knowing what label properties are being used to describe a particular entity. Semantic web data defines resources using URIs which are opaque and do not provide human reader-friendly detail about the resource. The data publisher makes use of label properties to provide a textual description of the resource and allows application developers to make use of these properties to learn more about the resource and/or use them to develop application interfaces.

As mentioned in Sect. 5.4.4, two types of label properties are defined in the EMP-AF: formal labels which are part of the W3C-based vocabularies, and domain labels which are defined by the particular ontologies. Figure 8.20 lists the query which retrieves the use of the domain labels and the formal labels used by data publishers to provide a textual description for the entities.

The U Ontology provides the `FormalLabel` and `DomainLabel` concepts to capture the label properties used in the dataset which are subclasses of `LabelUsage`. In Fig. 8.20, the query retrieves all the labels used for the instances of the BusinessEntity concept of the GoodRelations ontology. The names of the label properties are

Fig. 8.21 Result of
SPARQL query shown in
Fig. 8.20

flname	dlname
rdfs:label	gr:legalName
rdfs:comment	gr:category

displayed. The `LabelUsage` with RDFS entailment rules can be used to access the same result (similar to Req. 1) even if a distinction is not required.

Figure 8.21 shows the use of label properties to provide a textual description about the entity for human readability or user interface. Entities of the type BusinessEntity are described using both the domain and formal labels, which include `rdfs:label`, `rdfs:comment`, `gr:legalName` and `gr:category`. Application developers and data publishers can use this information to develop the user interface and provide textual description in the place of opaque URIs.

8.5.4.4 Req. 4: What Knowledge Patterns Are Available in the Dataset?

In this requirement, the user is interested in knowing the knowledge patterns that are prevalent in the published Semantic Web data. The knowledge patterns provide terminological knowledge in the sequence of paths comprising different path steps. Each path step is of a concept-predicate-concept pattern and different path steps which are linked (chained) in the RDF graph constitute a path.

The EMP-AF implements the technique and method for extracting the knowledge patterns present in the dataset to allow users to understand the prevalent structure of the schema level graph. The U Ontology represents all the conceptual elements to capture the components of the knowledge pattern, as depicted in Fig. 7.9.

Figure 8.22 lists the query which displays the knowledge patterns found in the dataset. The U Ontology implements the knowledge pattern conceptual model to allow users to represent and access the schema level triples included in it. The query displays all the schema level triples constituting the knowledge pattern.

8.5.5 Discussion on Findings

In this section, the usability and adequacy of the results from the EMP-AF is presented. To demonstrate the utilization, use cases are presented to reflect the common requirements of different types of user. Each use case represents the frequently occurring requirements from the perspective of different users. For each requirement, the method, technique and metrics implemented as part of the EMP-AF is discussed. In terms of allowing the user to make use of the output of empirical analysis, the

```
1   PREFIX uo: <http://example.uontology.org/v1#>
2
3   SELECT   ?kpsubname,   ?kpprename,   ?kpobjname
4   WHERE {
5           ?ontology  a                      uo:OntologyUsage.
6           ?ontology  uo:hasKnowledgePattern ?kp.
7           ?kp        uo:hasPath             ?path.
8           ?path      uo:hasPathStep         ?pathstep.
9           ?pathstep  uo:hasSubjectInPath    ?kpsub;
10                     uo:hasPropertyInPath   ?kppre;
11                     uo:hasObjectInPath     ?kpobj.
12          ?kpsub     uo:name                ?kpsubname.
13          ?kppre     uo:name                ?kpprename.
14          ?kpobj     uo:name                ?kpobjname.
15  }
```

Fig. 8.22 SPARQL query to display the knowledge patterns in the dataset

U Ontology is queried against each requirement. Based on the use cases and the solution offered by the EMP-AF and U Ontology, the findings are summarized in the following points:

1. The EMP-AF implements the methods and techniques which are required to empirically analyse the use of ontologies. The implementation of the framework has successfully addressed the requirements of the use cases related to the investigation phase of the OUSAF. Therefore, the **EMP-AF, its techniques and methods are effective in providing the required insight about the empirical analysis of ontology usage**.
2. The U Ontology is able to represent and capture the concepts pertaining to the Concept Usage Template, labelling and knowledge patterns. **The information required for each use case was retrieved by accessing the U Ontology, therefore the conceptual model formalizing the domain of ontology usage is considered adequate for addressing users requirements**.

8.6 Utilization of the Quantitative Analysis Framework (QUA-AF)

8.6.1 Aim

As mentioned in the previous section, the aim of the investigation phase is to empirically and quantitatively analyse the use of ontologies. The focus of this section is quantitative analysis, in which using the identified aspects, key dimensions which are important for measuring ontology usage, are defined. The quantitative analysis is performed from three dimensions: richness, technology and business. Based on these dimensions, the use of an ontology and its components are quantitatively measured, and a quantified rank of each term is obtained using a ranking approach. To quantitatively measure ontology usage, the QUA-AF is used.

8.6.2 Flow of Activities

The QUA-AF comprises three phases: data collection, computation and application. In QUA-AF, the dataset is analysed from three dimensions and each dimension requires a different type of dataset. For richness, the formalized ontological model and same dataset described in Sect. 8.3 is used. Separate data comprising semantic mark-ups that are supported by search engines is collected for the business dimension. The methods, techniques and metrics developed for the QUA-AF are then used to develop the web schema which provides a snapshot of the terminological knowledge that is published on the Web in a specific application area.

8.6.3 Use Cases

Quantitative analysis performed using the QUA-AF allows users to analyse ontology usage from different dimensions while providing a consolidated rank of terms. Each dimension is measured and captured independent of others to allow users to access information specific to their requirements.

To demonstrate the utilization of the QUA-AF, use cases are defined to reflect the usage scenarios applicable to different types of user.

Ontology developer's requirements
Req. (1) What is the richness value of the concepts in a given ontology?

Data consumer's requirements
Req. (2) Display the ontology terms based on their usage ranking?

Data publishers' requirements
Req. (3) List the terms that are being recognized by search engines?
Req. (4) What ontologies are being used in a given application area?
These requirements are used to analyse the utilization of the QUA-AF.

8.6.4 Utilization Analysis

8.6.4.1 Req. 1: What is the Richness Value of the Concepts in a Given Ontology?

In this requirement, the user is interested in knowing how the concept in a given ontology is structured. Structure refers to the typological characteristics being defined

```
1    PREFIX uo: <http://example.uontology.org/v1#>
2
3    SELECT ?conname ?richvalue
4    WHERE {
5            ?ontology   rdf:type              uo:OntologyUsage;
6                        uo:prefix             "gr".
7            ?ontology   uo:hasConceptUsage    ?oconcept;
8                        uo:name               ?conname.
9            ?oconcept   uo:hasRichness        ?ocrich.
10           ?ocrich     up:value              ?richvalue.
11   }ORDER BY DESC (?richvalue)
```

Fig. 8.23 SPARQL query to display the concepts and their richness value of a specific ontology

in the ontology to describe a concept. A concept is described by creating a type relationship with other concepts and specifying the attributes to capture the factual state of the entity conceptualized by the concept. The richness value, as mentioned in Sect. 6.3.1, quantifies the structural and typological characteristics of the concept, relationships and attributes.

The QUA-AF defines the metrics to measure the value of concepts and the ontology's authoritative documents are accessed by the framework to measure the richness value of ontology components. For each type of ontology component, respective metrics are defined to measure the richness value of concepts, relationships and attributes. These values are then represented in the U Ontology to allow users to access the quantitative analysis of the ontology usage.

Figure 8.23 displays the SPARQL query to list all the concepts of an ontology and their richness value. The list contains the concepts which have been used on the Web and their richness value is computed using the metrics defined as part of the QUA-AF (Sect. 6.3.1). The U Ontology captures the concepts and attributes which are necessary to represent the richness value of the ontology components, including relationships and their attributes. In the above query, the concepts of a given ontology (i.e. GoodRelations, which has "gr" as a prefix in the triple store) are accessed along with their richness value. The query displays the name and the value of the concepts in descending order.

Figure 8.24 displays the different concepts of the GoodRelations ontology with their concept richness values. This illustrates how the entities are conceptualized and semantically described in their formalized model. The concept richness value is combined with other metrics to generate a ranked list of ontology terms to allow users to use the terms based on their requirements (e.g., terms with a higher usage or richness value).

8.6.4.2   Req. 2: Display the Ontology Terms Based on Their Usage Ranking?

In this requirement, the user is interested in obtaining a list of terms based on their usage ranking. This includes all the concepts, relationships, and attributes defined

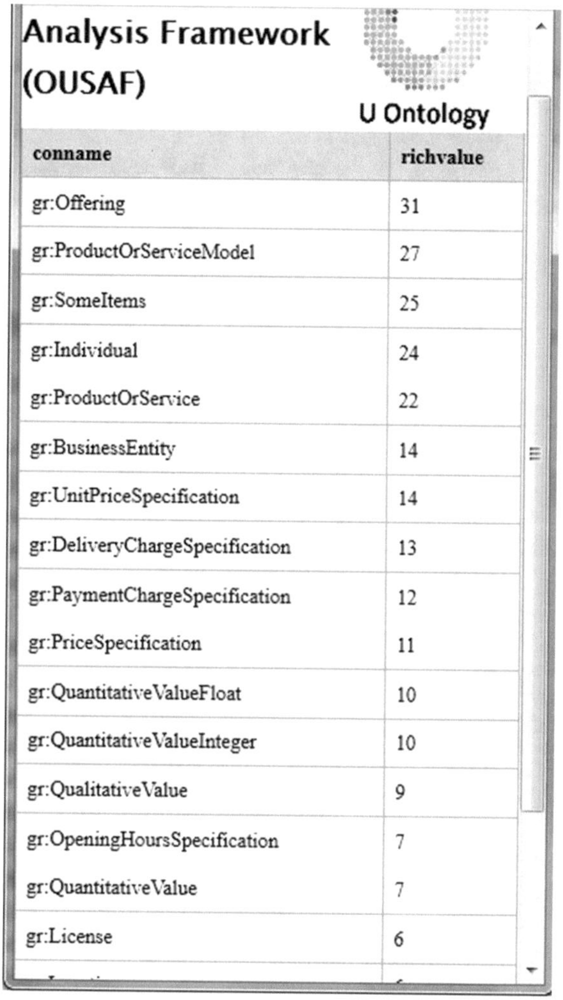

Fig. 8.24 Result of
SPARQL query shown in
Fig. 8.23

Fig. 8.24 Result of SPARQL query shown in Fig. 8.23

conname	richvalue
gr:Offering	31
gr:ProductOrServiceModel	27
gr:SomeItems	25
gr:Individual	24
gr:ProductOrService	22
gr:BusinessEntity	14
gr:UnitPriceSpecification	14
gr:DeliveryChargeSpecification	13
gr:PaymentChargeSpecification	12
gr:PriceSpecification	11
gr:QuantitativeValueFloat	10
gr:QuantitativeValueInteger	10
gr:QualitativeValue	9
gr:OpeningHoursSpecification	7
gr:QuantitativeValue	7
gr:License	6

by the ontology which have been used by data publishers. In the QUA-AF, metrics
for each dimension are defined and computed using their respective repositories, as
each requires different types of data for computation. To obtain the consolidated
rank comprising these three dimensions, the QUA-AF computes and consolidates
the values based on the ranking approach (Eq. (6.9)) presented in Sect. 6. Since the
priority or relevance of each dimension is controlled through weights specified by
the user, the U Ontology does not capture the consolidated ranking value. However,
it defines the concepts to represent the value computed for each dimension to allow
users to obtain the terms of each dimension and their values.

Figure 8.25 displays the terms of the ontologies and their values computed from
three dimensions. The query retrieves all the concepts and their usage, richness and

```
 1  PREFIX uo:  <http://example.uontology.org/v1#>
 2
 3  SELECT *
 4  WHERE {
 5          ?ontology     uo:prefix      "gr".
 6          {
 7          ?ontology     uo:hasConceptUsage     ?con.
 8          ?con          uo:name                ?cname.
 9          ?con          uo:hasRichness         ?conrich.
10          ?con          uo:isIncentividesBy    ?sengine.
11
12          ?con          uo:hasUsageValue       ?cuvalue.
13          ?conrich      uo:hasRichnessValue    ?crvalue.
14          ?sengine      uo:hasIncentiveValue   ?civalue.
15          }
16          UNION
17          {
18          ?ontology     uo:hasRelationshipUsage  ?rel.
19          ?rel          uo:name                  ?rname.
20          ?rel          uo:isIncentivisedBy      ?rincv.
21          ?rel          uo:relationshipValue     ?rrich.
22
23          ?rel          uo:hasUsageValue         ?ruvalue.
24          ?rrich        uo:hasRichnessValue      ?rrvalue.
25          ?rincv        uo:hasIncentiveValue     ?rivalue.
26          }
27          UNION
28          {
29          ?ontology     uo:hasAttributrUsage     ?att.
30          ?att          uo:name                  ?aname.
31          ?att          uo:isIncentivisedBy      ?attincv.
32          ?att          uo:attributeValue        ?attval.
33
34          ?att          uo:hasUsageValue         ?auvalue.
35          ?attval       uo:hasRichnessValue      ?arvalue.
36          ?attincv      uo:hasIncentiveValue     ?aivalue.
37          }
38  }LIMIT 50
```

Fig. 8.25 SPARQL query to display the usage of given ontology terms

incentive values computed by the QUA-AF that will be published through the U Ontology. However, the query is not able to compute the final consolidated rank value for each component since weights are required for this. The computation of the final rank value can be computed, based on the data provided by the query and by specifying the weights.

Figure 8.26 displays the list of ontology terms along with their usage, richness and incentive measures. These values help ontology owners and data publishers to analyse the usage from different dimensions based on their requirements, and by applying a threshold value (filter), terms with a certain usage or rank value can be obtained.

Concept	cuvalue	crvalue	civalue
gr:Offering	27165	31	3
gr:ProductOrServiceModel	6275	27	1
gr:SomeItems	12465	25	2
gr:Individual	32	24	1
gr:ProductOrService	8	22	2
gr:BusinessEntity	1714	14	3
gr:UnitPriceSpecification	14265	14	0
gr:DeliveryChargeSpecification	6	13	0
gr:PaymentChargeSpecification	15	12	0
gr:PriceSpecification	101	11	0
gr:QuantitativeValueFloat	6603	10	0
gr:QuantitativeValueInteger	92	10	0
gr:QualitativeValue	950	9	0
gr:OpeningHoursSpecification	703	7	0
gr:QuantitativeValue	0	7	0
gr:License	5	6	2
gr:Location	1057	6	2

Fig. 8.26   Result of SPARQL query shown in Fig. 8.25

8.6.4.3   Req. 3: List the Terms That Are Being Recognized by Search Engines?

In this requirement, the user, particularly a data publisher, is interested in knowing which terms are currently recognized by search engines. In the QUA-AF, commercial incentives for the business dimension are measured by identifying which terms are recognized by search engines when used to semantically describe information published on the Web. For this, three search engines are used and if the term is recognized by any of them, it is represented in the U Ontology. U Ontology provides the concept and attributes to allow users to access incentive-related information.

Figure 8.27 shows the query and Fig. 8.28 displays the terms that are recognized by search engines. This list helps data publishers to know which terms to consider if the user is interested in greater visibility of information on the Web. Similarly, it lets data publishers and application developers know which terms to prefer over others, based on their recognition by other search engines.

8.6.4.4   Req. 4: What Ontologies Are Being Used in a Given Application Area?

In this requirement, the user is interested to know which ontologies should be considered for information relevant to their domain. Several ontologies which have a

```
1   PREFIX uo: <http://example.uontology.org/v1#>
2
3   SELECT ?comname, ?sename
4   WHERE {
5    ?component      uo:isIncentivisedBy    ?se.
6    ?component      uo:name                ?comname.
7    ?se             uo:searchEngineName    ?sename.
8
9   }LIMIT 50
```

Fig. 8.27 SPARQL query to list the terms that are recognized by search engines

Fig. 8.28 Result of
SPARQL query shown in
Fig. 8.27

(OUSAF)	U Ontology
comname	sename
Address	Google
author	Google
locality	Google
Offer	Google
Organization	Google
Person	Google
photo	Google
postal-code	Google
Product	Google
Recipe	Google
Review	Google
street-address	Google
title	Google

high level of usage have been published on the Web, therefore it is more beneficial
to consider these. A detailed usage analysis of each ontology can be obtained, but
initially, it is important to know which ontology to consider. In several previous
requirements, ontology-specific information has been queried from the U Ontology;
however, in this requirement, the name of an ontology relevant to an application area
is required.

```
1   PREFIX uo: <http://example.uontology.org/v1#>
2
3   SELECT ?oname, ?oprefix
4   WHERE {
5       ?oua    rdf:type              uo:OntologyUsageAnalysis.
6       ?oua    uo:analysisOntology   ?ou.
7       ?ou     uo:name               ?oname.
8       ?ou     uo:prefix             ?oprefix.
9       ?ou     uo:isUsedby           ?ds.
10      ?ds     uo:industry           ?ind.
11
12      FILTER regex(str(?ind), "eCommerce", "i")
13  }
```

Fig. 8.29 SPARQL query to list the ontologies being used in a given application area

Figure 8.29 displays the name and prefix of the ontologies that have been used in a dataset and are relevant to the e-Commerce domain. The U Ontology captures the meta data regarding the dataset and the usage analysis to allow users to obtain information about when the analysis was performed, the nature of the data included in the dataset and the industry which the majority of the data represent. In the query, all the ontologies which have been identified and analysed by the OUSAF are retrieved, and using a filter, only ontologies labelled (tagged) as e-Commerce are displayed. The same query can be used to obtain a list of ontologies relevant or applicable to other application areas.

Figure 8.30 lists all the ontologies/vocabularies that are being used to describe information in a particular domain. The query retrieves the ontologies being used in e-Commerce domains and this information helps data publishers and application developers to know which ontologies to consider, whether for developing ontology-driven application or for using to semantically describe information on the Web.

8.6.5 Discussion on Findings

In this section, the usability and adequacy of the results from the QUA-AF are presented. To demonstrate the utilization, use cases are presented to reflect the common requirements of different types of users. For each use case, requirement analysis is presented and the methods, techniques and metrics implemented as part of the QUA-AF are discussed. For each user, SPARQL queries are made to the U Ontology, which represents the usage analysis-related data, to provide the information needed by the user. Based on the use cases and solutions offered by the QUA-AF and the U Ontology, the findings are summarized in the following points:

1. The QUA-AF implements the methods and techniques which are required to quantitatively analyse the use of ontologies. Through the implementation of the framework, it has successfully addressed the requirements of the use cases related

Fig. 8.30 Result of SPARQL query shown in Fig. 8.29

(OUSAF)	U Ontology
oname	**oprefix**
GoodRelations	gr
Friend of a Friend	foaf
vCard	vcard
vocabulary (google)	v
open graph protocol (facebook)	og
Dublin Core	dc
review	rev
yahoo	yahoo
product ontology	pto
geo	geo

to the investigation phase of the OUSAF. Therefore, the **QUA-AF is capable and its techniques and methods are satisfactorily able to provide the required insight into the quantitative measures regarding ontology usage**.

2. The U Ontology is able to defines concepts pertaining to the richness, technology and business dimensions and the quantified measures obtained through the metrics defined in the QUA-AF. **The information required for each use case was retrieved by accessing the U Ontology, therefore the conceptual model formalizing the domain of ontology usage is considered adequate for addressing users requirements**.

In the next section, the achievements obtained through the utilization phase are discussed.

8.7 Benefits of the Utilization Phase

The objective of the utilization phase is to enable users to access the OUSAF and analyse ontology usage, based on their requirements. The communication between users and the computational components of the OUSAF is facilitated through the use of the U Ontology. The benefits for each type of user obtained in the utilization phase are summarized in the following subsections.

8.7.1 Benefits from an Ontology Developer's Perspective

An ontology developer is interested in knowing the performance of a given ontology in terms of its usage. By using the OUSAF, ontology developers can determine:

- the usage of a given ontology, which includes the number of data sources that are using the ontology, the number of instances created using the ontology namespace and the ontology meta- information. In Sect. 8.5.4.1, the request that is made to the OUSAF by posing the query is shown in Fig. 8.6. The query is able to provide the required information to the user who can then use it for further processing.
- the usage of different components of an ontology. The components are the concepts, object properties and attributes which have instantiation on the Web. In Sect. 8.6.4.1, the requirement of obtaining the ontology adoption level is achieved by posing the SPARQL query shown in Fig. 8.15, which provides a list of the names of the ontology components which have been used on the Web.

8.7.2 Benefits from a Data Consumer's Perspective

A data consumer is interested in knowing not only which ontologies are available for use but also exactly what is being used in these ontologies. By using the OUSAF, the data consumer can:

- obtain a list of all ontologies which are currently being used by other publishers to describe information related to the domain of interest. In Sect. 8.5.4.3, the requirements for retrieving a list of ontologies used in a specific domain are obtained through the query shown in Fig. 8.10. This query provides the name, prefix of the ontologies and also the number of different data sources who are using it. The availability of this information helps data consumers to achieve their objectives.
- determine what labelling properties are being used by the community and are provided by the ontologies to textually describe the information. The textual description provided by these labels allows human readers to understand the entity and also allows user interfaces to display labels rather than opaque URIs. In Sect. 8.6.4.3, the requirements for retrieving the list of label properties used by publishers is obtained through the query shown in Fig. 8.20. This query is capable of providing details of label properties such as their name and the ontology to which they belong.

8.7.3 Benefits from a Data Publisher's Perspective

Data publishers prefer to reuse ontologies to benefit from the advantage of the existing support and acceptance in their respective communities. To do this, they need to learn about the current usage level of various ontologies and their co-usability with other ontologies. The benefits available to data publishers through the utilization of the OUSAF are described below.

- One of the new motivations for using ontologies on the Web is the support they are given by search engines. For data publishers, it is important to know which terms are recognized by which search engines, as this will help them to improve the visibility of their information on the Web. In Sect. 8.6.4.3, the list of terms being supported by search engines and the names of the search engines which support them are obtained through the query shown in Fig. 8.27. Using the results obtained through this query helps data publishers to choose the terms they should use for their semantic description.
- In Sect. 8.4.4.5, the cohesive groups of ontologies with similar usage are obtained through the query presented in Fig. 8.13. The identification and availability of cohesive groups helps data publishers to understand the prevalent semantic structure available in the currently published Semantic Web data, and to consider it for their semantic annotation.

Based on the above-mentioned benefits of the OUSAF and the use cases in Sects. 8.5–8.7, it can be concluded that the OUSAF, through its computational components (OUN-AF, EMP-AF, and QUA-AF) and the U Ontology, successfully demonstrates the utilization of the proposed framework.

8.8   Recapitulation

In this chapter, the utilization phase of the OUSAF was presented. To demonstrate the utilization of the OUSAF, an approach was presented which accesses each solution component, i.e. OUN-AF, EMP-AF, and QUA-AF from different users requirement perspectives. For each type of user, different use cases were presented to obtain the required information from the solution components. The results obtained in each use case were then analysed to see whether the information was useful and suitable for the user.

In the next chapter, the evaluation of the U Ontology is presented.

Chapter 9
Evaluation of U Ontology

9.1 Introduction

The utilization of the OUSAF and its computational components were presented in the previous chapter. The U Ontology, which conceptualizes the Ontology Usage Analysis domain, was used to obtain ontology usage-related information from the OUSAF. In this chapter, the U Ontology will be evaluated using an ontology evaluation methodology to measure the quality of the developed ontology. For an ontology to be of good quality and remain useful for its users, it needs to conform to a set of good practices. These practices are analysed using evaluation techniques which assess an ontology based on specific criteria to ensure that the developed ontology meets the user's expectations.

The ontology evaluation methodology and observations on its performance are presented in this chapter. The criteria used by the adopted methodology to evaluate the various aspects of the U Ontology are described in Sect. 9.2. In Sects. 9.3–9.8, the U Ontology is evaluated using the methods and aspects supported by the methodology. A summary of the U Ontology evaluation is presented in Sect. 9.9. Section 9.10 concludes the chapter.

9.2 Methodology for Ontology Evaluation

The purpose of ontology evaluation and its role in ontology engineering was discussed in Chaps. 1 and 2. There are number of frameworks (Sect. 2.3.2 discusses these in detail) for ontology evaluation, all of which have the common objective of assessing the quality of a given ontology. While all the evaluation frameworks attempt to answer the question of how to assess the quality of an ontology for the Web, they differ in their approaches and techniques. References [40, 280] classified the ontology evaluation approaches into the following categories:

© Springer International Publishing AG 2018
J. Ashraf et al., *Measuring and Analysing the Use of Ontologies*, Studies
in Computational Intelligence 767, https://doi.org/10.1007/978-3-319-75681-3_9

- Self evaluation of ontologies. In this category, the golden standard approach is adopted in which an ontology is assessed by comparing it with another ontology, e.g. [184].
- Context-dependent evaluation of ontologies. The context is often specified by the inclusion of the additional artefact used to develop the ontology. The competency question also specifies the context of the ontology, e.g [125].
- Application specific evaluation of ontologies. This means evaluating an ontology by using it within an application. This is also known as application-based ontology evaluation, e.g. [40].
- Application and task specific evaluation of ontologies. These approaches are also known as task-based ontology evaluation, e.g. [220].

While each of the above categories has a different approach, each gains from the evaluation of the other categories. This means that the problems identified and rectified by the technique of one category will benefit the approaches in other categories.

The methodology adopted in this book for the evaluation of the U Ontology is proposed by [280] and also discussed in [240, 276, 283]. The authors' proposed approach is based on the premise that a single measure to assess the quality of an ontology is elusive, and deriving concrete measures to identify the errors and loopholes in ontologies is a more practical approach. They state that:

> [...] instead of aiming for evaluation methods that tells us if an ontology is good, we settle for the goal of finding ontology evaluation methods that tell us if an ontology is bad, and if so, in which way.

Following this approach, the evaluation methodology provides a set of techniques and methods to evaluate an ontology from several aspects and helps to determine whether the ontology is satisfactory or unsatisfactory.

In the next subsections, we discuss the criteria which should be considered in evaluating the quality of an ontology.

9.2.1 Criteria for Ontology Evaluation

In his dissertation, Vrandecic [280] discusses the various criteria which a good ontology should meet. These criteria are briefly described below. Several methods are proposed under each criterion which will be used to assess the U Ontology.

- *Accuracy*: Accuracy is a criterion that states that the knowledge represented by the ontology, including the axiomatic triples, conforms to the knowledge of the stakeholders about the domain [280]. A higher accuracy value is achieved with the correct description of ontology components, which include classes, properties, and individuals.
- *Adaptability*: Adaptability measures how flexible an ontology is in addressing user needs [280]. Since ontologies are meant to be used on the Web, and their Web usage cannot be predicted, the conceptual foundation should at least be capable of fulfilling the range of anticipated tasks.

- *Clarity*: Clarity measures how effectively the ontology provides the understanding and meanings of the terms defined in the ontology [280]. As a best practice, the terms defined in the ontology to name classes, properties and individuals should be understandable and unambiguous. This means that the definition of terms should be independent of the context and easily interpreted by users.
- *Completeness*: This measures how well an ontology covers the domain of interest [280]. The requirements should be answered within the scope of the ontology and address the following issues of completeness:

  - Completeness with regard to language (is the textual description required for a task reasonably detailed?)
  - Completeness with regard to domain (are all the key concepts representing the entities of the domain covered? Is it possible to represent all the individuals by the concepts?)
  - Completeness with regard to the application requirements (is all the data needed by the application present and representable by the ontology?)

- *Computational efficiency*: Computational efficiency measures the ability of the tools to work with the ontology [280]. These tools are: the databases (or triple stores) to store the ontology and individuals, reasoners to reason over the ontology based on the axioms implemented in the ontology (and RDFS/OWL), and query processing. Reasoners in particular are important because they are often used to infer entailed knowledge, query answering, classification, and consistency checking.
- *Conciseness*: Conciseness measures whether the ontology includes elements that are not relevant to the domain being represented through the ontology [280]. This includes the definition of concepts which are not directly related to the domain of interest, or the presence of concepts which give redundant representation of the semantics.
- *Consistency*: Consistency indicates whether the ontology excludes contradiction in the model [280]. Generally, consistency concerns logical consistency and coherence, and principles are defined to ensure that an ontology remains both consistent and coherent.
- *Organizational fitness*: This measures how easy or challenging it is for an ontology to be implemented in an organization [280]. It includes aspects such as people (acceptance, resistance to change), tools (development tools, databases, reasoners, software licences), and technology and methodology (familiarity with the technology used in ontologies and the methodologies that form part of the organizational information architecture).

9.2.2   Aspects to be Analysed for Ontology Evaluation

Vrandecic et al. [280] proposed six aspects which should be considered in evaluating ontologies using the above-mentioned criteria as follows: Vocabulary, Syntax, Structure, Semantics, Representation and Context, which are defined below.

- *Vocabulary*: This aspect refers to the names that are used in the ontology to describe the resources and literal values. The evaluation of the vocabulary aspect of the U Ontology is discussed in Sect. 9.3.
- *Syntax*: This aspect refers to the serialization format used to encode the ontology. There are several types of syntax available and different ontologies use different syntaxes, but all of them generate a graph. The evaluation of the syntactical aspect of the U Ontology is discussed in Sect. 9.4.
- *Structure*: This aspect represents how an ontology graph is arranged. Even though all ontologies are based on the RDF graph model, they can vary structurally. The evaluation of the structural aspect of the U Ontology is discussed in Sect. 9.5.
- *Semantics*: This aspect is about the formal meaning being represented by the ontology. The evaluation of the semantics aspect of the U Ontology is discussed in Sect. 9.6.
- *Representation*: This aspect represents the relationship between structure and semantics. The evaluation of the representational aspect of the U Ontology is discussed in Sect. 9.7.
- *Context*: This aspect covers the features of an ontology when compared with other artefacts. The evaluation of the contextual aspect of the U Ontology is discussed in Sect. 9.8.

Twenty-three methods are proposed in [280] to facilitate the evaluation of ontologies from these six aspects. These methods will be used to evaluate the U Ontology from the above-mentioned six aspects.

The applicable methods for the evaluation of the U Ontology from each aspect are presented with their definition (reproduced from [280]), brief description and evaluation result.

9.2.3   Metrics to Quantify the Evaluation Findings

Each method is applied using the applicable procedure, technique or process suggested by the methodology [280] and the results of each method are used to evaluate the U Ontology. The evaluation results obtained for each method are descriptive in nature and need to be analysed and interpreted, keeping in view the ontology and the knowledge base comprising instance data (i.e. the populated U Ontology). While these results help in understanding the quality of the ontology, they do not quantify the ontology's overall performance. To quantify the U Ontology evaluation and provide conclusive remarks about the results, the following four metrics are used.

- **Verified**: The method is applied as required and the evaluation results obtained are positive. Positive indicates that no problem is found and the results are as expected.
- **Not Applicable**: The method is not applicable to the ontology. This could be because the given method cannot be computed due to the ontology language or limitations of the reasoners.
- **Deferred**: The method is applicable but cannot be verified due to technical or time constraints and will be considered in future work.
- **Failed**: The method was applied but did not achieve the expected results.

At the end of the evaluation, the value of each metric is accumulated (from the evaluation of the methods of each aspect) and used to summarize the evaluation of the U Ontology.

9.3   U Ontology Evaluation: Vocabulary Aspect

This aspect evaluates the vocabulary of the ontology. The vocabulary of an ontology is the set of all the names used to define the terms (components of an ontology). Names can be either URIs or literals. URI references identify resources and thus provide unique identifiers for all the ontology components, whereas literals are names that are mapped to a concrete data value. In addition to URIs and literals, ontologies have unnamed entities known as blank nodes. The methods applicable to vocabulary and their evaluation in the U Ontology are presented in the following subsections.

9.3.1   Method 1: Check the Protocols Used

This method is used to check the protocols used in the ontology. The definition of the method is as follows:

> **Method 1 (Check used protocols) [280]**
> "All URIs in the ontology are checked to ensure they are well-formed URIs. The evaluator has to choose a set of permitted protocols for the evaluation task. The usage of any protocol other than HTTP should be explained. All URIs in the ontologies have to use one of the permissible protocols."

Most names in ontologies are URI references (generic form of URLs) [23]. URI references are strings that start with protocols. The recommended protocol for the URIs is HTTP because this allows applications (or even ontologies) to resolve the URIs. Resolving the URI means that more information about the identified resource is provided. The Linked Data principle recommends using the HTTP protocol in URIs for dereferencing.

```
1
2    <owl:Class rdf:about="http://oua.uontology.org/v1#FormalLabel">
3         <rdfs:subClassOf rdf:resource="http://oua.uontology.org/v1#LabelUsage"/>
4    </owl:Class>
5
6    <owl:ObjectProperty rdf:about="http://oua.uontology.org/v1#hasRichness">
7      <rdfs:range  rdf:resource="http://oua.uontology.org/v1#ConceptRichness"/>
8      <rdfs:domain rdf:resource="http://oua.uontology.org/v1#ConceptUsage"/>
9      <rdfs:subPropertyOf rdf:resource="&owl;topObjectProperty"/>
10   </owl:ObjectProperty>
11
```

Fig. 9.1 The URI and HTTP protocol used in the U ontology

Evaluation: In the U Ontology, all the URIs use the HTTP protocol and thus are resolvable. Each URI can be dereferenced and provides textual description for human readability.

Figure 9.1 shows a snippet of the U Ontology in which it can be seen that URI references (i.e. `http://oua.iontology.org/v1#FormalLabel`; Line 2) for the terms (e.g. `FormalLabel` class and `hasRichness` property) use the HTTP protocol.

Conclusion: `Verified`

9.3.2 Method 2: Check Response Codes

This method checks the response code of the HTTP request. The definition of the method is as follows:

Method 2 (Check response codes) [280]
"For all HTTP URIs, make a HEAD call (or GET call) on them. The response code should be 200 OK or 303 See Other. Names with the same slash namespace should return the same response codes, otherwise an error is indicated."

Resolving an `HTTP URI` reference returns an `HTTP` response code along with the content related to the referenced URI. There is a predefined set of codes with special meanings to interpret the codes, and the appropriate code should be provided upon the `HTTP GET` request. There are two types of resources on the Web: information resources and non-information resources [29].

- Information resources: When a URI links to an information resource, then the client receives a snapshot of the resource's current state by the HTTP response code 200 OK. This snapshot is generated by the URI owner's server.
- Non-information resources: Such resources cannot be directly dereferenced. So in this case, the server instead of sending a snapshot of the resource's current state will send to the client the URI of the resource using the HTTP response code 303

See Other. This is called a 303 redirect. The client by dereferencing this URI will get a representation that describes the non-information resource. Non-information resources cannot be dereferenced directly, therefore Web ar

Evaluation: In the deployment of the U Ontology, the server is configured to provide both HTTP Response 200 OK (for information resources) and HTTP Response code 303 See Other (for non-information resources). The content negotiation for non-information resources is also possible; however the hash approach (URI dereferencing) discussed in the next method is followed.

Note: The current version of the ontology is deployed on the intranet. In the future when our internet server is ready, the ontology will be moved to a live server.

Conclusion: `Verified`

9.3.3   Method 3: Look up Names

This method checks whether the Hash to Slash approach for the URI reference is used in the ontology.

Method 3 (Look up names) [280]

"For every name that has a hash namespace, make a GET call against the namespace. For every name that has a slash namespace, make a GET call against the name. The content type should be set correctly. Resolve redirects, if any. If the returned resource is an ontology, check whether the ontology describes the name. If it does, N is a linked data conformant name. If it does not, the name may be wrong."

HTTP URIs need to be capable of being dereferenced, so that the HTTP client can look up the URI to retrieve the description of a resource. A URI identifying a real world object is different from a URI referring to a document describing the real world object. To avoid ambiguity, two separate URIs are used to identify objects. Content negotiation is used to provide the HTML for humans and RDF for machines when a URI about a resource is dereferenced. There are two strategies for naming non-information resources: 303 URI and hash URI. For example, for a non-information resource "Book", the 303 URI will be http://example.org/ontology/Book, and has a URI of http://example.org/ontology#Thesis. Both approaches have advantages and disadvantages which should be considered when deciding on the appropriate approach.

Evaluation: The U Ontology uses HTTP URI Reference to name the resources (terms). In the implementation of an ontology, the hash URI approach is adopted based on the recommendation suggested in [26, 140, 179]. The advantages of using URI is that it is downloaded in one pass because when the hash URI is looked up, only the namespace is resolved and the fragment identifier is not sent to the server. This approach is adopted on the basis that the ontology size is small and will not be frequently changed, and instance data will not increase quickly and frequently. A disadvantage of using the hash URI approach is that it can lead to large amounts of data being unnecessarily transmitted to the client if the namespace description (ontology) consists of a large number of triples. In the instance described, this disadvantage is therefore not applicable, at least in the near future. The advantage of using hash URI is the avoidance of unnecessary trips to the server (courtesy of content negotiation). In Fig. 9.1, lines 6–9 show that hash URIs are used for the names of resources.

Conclusion: `Verified`

9.3.4 Method 4: Check Naming Conventions

This method checks the naming convention used to name terms in an ontology. The definition of the method is as follows:

> **Method 4 (Check naming conventions) [280]**
>
> "Correct naming can be checked by comparing the local part of the URI with the label given to the entity, or by using lexical resources like Wordnet [94]. Formalize naming conventions (like multi-word names and capitalization) and test whether the convention is applied to all the names of a namespace. Check whether the URI fulfils the general guidelines for good URIs, i.e. check length, inclusion of query parameters, file extensions, depth of directory hierarchy, etc. Note that only local names from the same namespace, not all local names in the ontology, need to consistently use the same naming convention, i.e. names reused from other ontologies can use different naming conventions."

The URI standard [24] states that the URI should be treated as opaque and no formal meanings should be associated with the URIs, aside from using the appropriate protocol. However, the Semantic Web community and Linked Data best practices recommend using meaningful URIs which invoke certain denotation to help human users make sense of the URI. Using a pure opaque URI such as `http://example.org/abd#1234` provides no clue to the resource it identifies, whereas `http://example.org/student/JamshaidAshraf` provides an indication that this resource concerns a person. There are naming conventions which a URI should follow to allow human users to establish an understanding of the resource denoted by the URI. There should be consistency in naming resources, and common practices should be adopted wherever possible. For example, using camel casing for multi-word names (e.g `JamshaidAsharf` instead of `Jamshaid_ashraf` or `jamshaidashraf`).

Evaluation: The names used in the U Ontology follow the naming convention recommended in [140, 229]. The following conventions are adopted:

- The name used in the URI to denote the resource is closely matched to the labels given to the entity. This increases human readability and in the case where a consuming application uses it in the interface, the URI still communicates clues about the entity.

```
1   . . . . . . . . . . . . . . . . . .
2   <!DOCTYPE rdf:RDF [
3       <!ENTITY uo "http://oua.uontology.org/v1#" >
4       <!ENTITY owl "http://www.w3.org/2002/07/owl#" >
5       <!ENTITY skos "http://www.w3.org/2004/02/skos-core#" >
6       <!ENTITY rdfs "http://www.w3.org/2000/01/rdf-schema#" >
7   ]>
8   . . . . . . . . . . . . . . . . . .
9   <rdf:RDF xmlns="http://oua.uontology.org/v1#"
10       xml:base="http://oua.uontology.org/v1"
11       xmlns:rdfs="http://www.w3.org/2000/01/rdf-schema#"
12       xmlns:owl="http://www.w3.org/2002/07/owl#"
13       xmlns:skos="http://www.w3.org/2004/02/skos-core#"
14       xmlns:uo="http://oua.uontology.org/v1#">
15       <owl:Ontology rdf:about="http://oua.uontology.org/v1#">
16           <owl:imports rdf:resource="http://omv.ontoware.org/2005/05/ontology"/>
17       </owl:Ontology>
18   . . . . . . . . . . . . . . . . . .
19   <!-- http://oua.uontology.org/v1#AttributeValue -->
20       <owl:Class rdf:about="&uo;AttributeValue"/>
21   . . . . . . . . . . . . . . . . . .
22   <!-- http://oua.uontology.org/v1#AtrributeUsage -->
23       <owl:Class rdf:about="&uo;AtrributeUsage">
24           <rdfs:subClassOf rdf:resource="&uo;Term"/>
25       </owl:Class>
26   . . . . . . . . . . . . . . . . . .
27       <!-- http://oua.uontology.org/v1#hasAttribute -->
28       <owl:ObjectProperty rdf:about="&uo;hasAttribute">
29           <rdfs:comment xml:lang="en">
30               This property allows to represent relationships present between
31               concepts based on their usage on real world implementations
32           </rdfs:comment>
33           <rdfs:range rdf:resource="&uo;Attribute"/>
34           <rdfs:domain rdf:resource="&uo;ConceptUsage"/>
35       </owl:ObjectProperty>
36   . . . . . . . . . . . . . . . . . .
37       <owl:DatatypeProperty rdf:about="&uo;searchEngineName">
38           <rdfs:domain rdf:resource="&uo;SearchEngine"/>
39           <rdfs:range rdf:resource="&rdfs;Literal"/>
40       </owl:DatatypeProperty>
```

Fig. 9.2 Excerpt from U ontology

- Camel casing is used for multi-word names
- URIs do not contain any meta-data, technology clues, or query parameters, as recommended in [23]
- Names for the class and properties are based on the naming convention followed in programming languages.

Figure 9.2 shows the excerpt taken from the U Ontology and will be used to demonstrate the implementation of a number of methods. For example, line 37 in Fig. 9.2 defines a name `searchEngineName` using the naming convention of camel casing, and line 34 displays the name of the concept comprising two words (i.e. `ConceptUsage`).

Conclusion: `Verified`

9.3.5   Method 5: Metrics of Ontology Reuse

This method checks the reusability factor adopted in the ontology.

> **Method 5 (Metrics of ontology reuse) [280]**
> "We define the following measures and metrics:
>
> - Number of namespaces used in the ontology N_{NS}
> - Number of unique URIs used in the ontology N_{UN}
> - Number of URI name references used in the ontology N_N (i.e. every mention of a URI counts)
> - Ratio of name references to unique names $R_{NU} = N_{UN}/N_N$
> - Ratio of unique URIs to namespaces $R_{UNS} = N_{UN}/N_{NS}$
>
> Check the following constraints. The percentages show the proportion of ontologies that fulfill this constraint within the Watson EA corpus, thus showing the probability that ontologies not fulfilling the constraint are outliers.
>
> $R_{NU} < 0 : 5 (79 : 6\%)$
> $R_{UNS} < 5 (90 : 3\%)$
> $N_{NS} >= 10 (75 : 0\%)$ "

The reuse of names (terms) is highly recommended in the Semantic Web because it eases the sharing, exchange and aggregation of information. The reuse of terms defined in other ontologies has been observed in the literature; for example, the terms defined in W3C-based vocabularies, terms in *foaf, dc, geonames* are reused by Semantic Web data publishers.[1]

Evaluation: In the U Ontology, terms from different ontologies are reused to take advantage of their adoption and built-in support in several tools. The vocabularies from which the terms are reused are the Ontology Metadata Vocabulary (OMV), Ontology Application Framework, Dublin Core (DC), FOAF, and vCard. Details of the reused ontologies and terms are presented in Sect. 8.4.4.

Conclusion: `Verified`

9.3.6   Method 6: Check Name Declaration

This method checks the presence of name declarations in the ontology

> **Method 6 (Check name declarations) [280]**
> "Check every URI to see whether a declaration of the URI exists. If it does, check whether the declared type is consistent with the usage. It is possible to detect erroneously introduced punning in this way."

---

[1] Updated statistics on the use of different terms in the LOD cloud are available at http://stats.lod2.eu/stats; retr. 20/11/2017.

Even though Web ontologies do not require names to be declared, declaring them is recommended. Declaring them in an ontology helps reasoners to decide whether a name which appears to match another name is a type or a new resource.

Evaluation: In the U Ontology, all the names are declared before they are used in the definition.

Line 20 in Fig. 9.2 declares the name of the `AttributeValue` class. Similarly, every URI that names the ontology components (class, relationship, and attribute) is explicitly declared before being used in the ontology.

Conclusion: `Verified`

9.3.7   Method 7: Check Literals and Data Type

This method checks whether a set of allowed data types is used.

> **Method 7 (Check literals and data types) [280]**
>
> "A set of allowed data types should be created. All data types beyond those recommended by the OWL specifications should be avoided. There should be a very strong reason for creating a custom data type. `xsd:integer` and `xsd:string` should be the preferred data types since they have to be implemented by all OWL conformant tools.
>
> Check whether the ontology uses only data types from the set of allowed data types. All typed literals must be syntactically valid with regard to their data type. The evaluation tool needs to be able to check the syntactical correctness of all allowed data types."

Literals are used to describe the states of the entities. Literals provide factual statements about the resource; for example, `Jamshaid Ashraf` is the author of this `Book`. So `Jamshaid Ashraf` is the literal value describing the author of the `Book` resource. Literals are typed in ontologies, which means that each literal has a data type associated with it which tells what data value is expected for that literal. Even though Semantic Web standards allow new custom data types to be defined, they should be avoided wherever possible.

Evaluation: In the U Ontology, no custom data type is used and only the data types which are part of the standard types are used. Most of the textual properties are of type `xsd:string` and numbers are of type `xsd:string`. The other data types which are used are `xsd:anyURI`, `xs:datatime`, and `xsd:int`.

Figure 9.2 shows the declaration of data properties and the data type from the allowed (recommended) data types. For example, in line 37, `searchEngineName` the data type properties are declared to be of type `rdfs:Literal`.

Conclusion: `Verified`

9.3.8   Method 8: Check Language Tag

This methods checks the use of language tags with literals.

> **Method 8 (Check language tags) [280]**
> "Check that all language tags are valid with regard to their specification. Check whether the shortest possible language tag has been used (i.e. remove redundant information such as restating default scripts or default regions). Check whether the stated language and script are actually the ones used in the literal.
> Check whether the literals are tagged consistently within the ontology. This can be checked by counting n_l, the number of occurrences of the language tag l occurring in the ontology. Roughly, n_l for all l should be the same. Outliers should be inspected."

Language tags can be used with plain literals (textual information) to tell tools which human language is used for literal values. The availability of different language tags, for example English and Arabic, allows tools to be displayed based on the users local preferred language.

Evaluation: In the present version of the U Ontology, only the @en (English) language tab is used for the textual description of entities, particularly for the values of rdfs:label and rdfs:comment properties. However, textual description in other languages, particularly German, French, Chinese and Arabic, is planned as part of future work.

In Fig. 9.2, the description of the hasAttribute property is given and the language tag (line 29) is used to let tools know in which natural language the description is provided.

Conclusion: Verified

9.3.9   Method 9: Check Labels and Comments

This method checks the use of label and comment properties for entities.

> **Method 9 (Check labels and comments) [280]**
> "Define the set of relevant languages for an ontology. Check whether all label and comment literals are language tagged. Check whether all entities have a label in all languages defined as being relevant. Check whether all entities that need a comment have one in all relevant languages. Check whether the labels and comments follow the style guide defined for the ontology."

Human readable names are provided for the entities (or terms) to allow humans to understand the ontology, its purpose and scope. As a recommended practice, all the terms defined in the ontology should have labels and comments in the appropriate language, marked with the language tag.

Evaluation: As mentioned in the previous method, the `rdfs:label` property is used to provide human readable names description of the term (entity) for each term defined in the U Ontology. However, only the English language tag is used. In future work, aside from providing textual descriptions in other languages, the use of other textual properties such as `skos:prefLabel` will be considered.

Conclusion: `Verified`

9.3.10 Method 10: Check for Superfluous Blank Nodes

This method checks the use of necessary and unnecessary blank nodes.

> **Method 10 (Check for superfluous blank nodes) [280]**
> "Tables 2.1 and 2.2 list all cases of structurally necessary blank nodes in RDF graphs. Check every blank node to see whether it belongs to one of these cases. Apart from these, no further blank nodes should appear in the RDF graph. All blank nodes which are not structurally necessary should be listed as potential errors."

Blank nodes are RDF features and are used to represent a node in the RDF graph without giving it an explicit name (that is, a URI). Such nodes can be internally referred but are not exposed to external applications.

Evaluation: In the U Ontology, no blank nodes are used and in future versions will be avoided if possible.

Conclusion: `Verified`

9.4 U Ontology Evaluation: Syntax Aspect

This aspect evaluates the syntax that is used to serialize the ontologies. Several serialization syntaxes are available, each with advantages and disadvantages. The other syntax-related issues discussed in this aspect are comment style, XML validation, and the creation of XML Schema.

9.4.1 Method 11: Validating Against an XML Schema

This method checks the syntax of the implemented ontology.

Method 11 (Validating against an XML schema) [280]

"An ontology can be validated using a standard XML validator under specific circumstances. In order to apply this, the ontology needs to be serialized using a pre-defined XML schema. The semantic difference between the serialized ontology and the original ontology will help to discover incompleteness of the data (by finding individuals that were in the original ontology but not in the serialized ontology). The peculiar advantage of this approach is that it can be used with well-known tools and expertise."

The conceptual model of an ontology is implemented using different serialization formats. Generally, there are two types of serialization syntax; one that describes a graph (e.g RDF/XML [17], N triple [120]), and another that describes the ontology directly (e.g. Manchester Syntax [153], OWL Abstract Syntax [219], or the OWL XML Syntax [152]). Two main observations are made pertaining to the syntactical approach followed in ontology serialization: (a) RDF-style comments should be used rather than XML-style comments; (b) prefixes (qualified names) which have already been adopted by the community to refer to the ontologies in ontology serialization should preferably be used. Additionally, the XML validation should be performed on an ontology to verify its conformance to the serialized syntax ontology on which it is built.

Evaluation: The U Ontology is serialized using RDF/XML syntax which is essentially based on XML Syntax. XML validation is performed to validate the syntax using the ontology development tool. Both XML style comments (e.g. `<!- - this is xml style - ->`) and RDF style comments (e.g. a triple with RDF comments (e.g. `<Jamshaid> rdfs:comment "Jamshaid is a Researcher"`) are used in the ontology. As pointed out in [280], XML style comments are often lost when two ontologies are merged or mapped, therefore RDF comments should be used. It is important to note that RDF specification as such does not have a standard way to provide in-line comments in the RDF document (similar to XML comments `<!- - text - ->`); however, the community has adopted the use of the "#" to mark the followed text in the line as a comment. In the U Ontology, "#" based comments were not used; only XML style comments were used and where possible `rdfs:comment` properties were used.

Conclusion: `Verified`

9.5   U Ontology Evaluation: Structural Aspect

This aspect evaluates the structural properties of the ontology graph. Ontologies are built using RDF, which is a graph model. To analyse the ontologies structurally, several measures are proposed which offer the following advantages:

• Structural measures are easy to compute from the ontology graph.
• The majority of tools provide support for structural measures.

- The structural measures are quantifiable and easy to interpret and visualize.
- The structural measures results are programmable, interchangeable and interoperable.

Different components of the ontology graph are observed and include but are not limited to subsumption hierarchy, semantic similarity and pattern recognition.

9.5.1   Method 12: Ontology Complexity

This method checks the structural characteristics of an ontology.

Method 12 (Ontology complexity) [280]
"We define measures by counting the appearance of each ontology language feature. We do this by first defining a filter function $O_T : O \rightarrow O$ with T being an axiom or an expression type. O_T returns all the axioms of axiom type T or all axioms having an expression of type T.
We also define a counting metric $N_T : O \rightarrow \mathbb{N}$ as $N_T(O) = |O_T(O)|$.
We also define $N(O) = |O|$.
We then further define a number of shortcuts, derived from the respective letters that define DL languages, for example:
- Number of subsumptions $N_{\sqsubseteq}(O) = N_{SubClassOf}(O) = |O_{SubClassOf}(O)|$: the number of subsumption axioms in the ontology
- Number of transitives $N_+(O) = N_{TransitiveProperty}(O)$: the number of properties being described as transitive
- Number of nominals $N_O(O) = N_{OneOf}(O)$: the number of axioms using a nominal expression
- Number of unions $N_{\sqcup}(O) = N_{UnionOf}(O)$: the number of axioms using a union class expression

With these numbers we can use a look-up tool such as the description logics complexity navigator. If $N_O > 0$, then the nominals feature has to be selected; if $N_+ > 0$ we need to select role transitivity, and so on. The navigator will then give us the complexity of the used language fragment (as far as known).
We further define $H(O) : O \rightarrow O$ as the function that returns only *simple subsumptions* in O, i.e. only those SubClassOf axioms that connect two simple class names."

Structural measures obtained from the ontology graph provide a number of interesting features, such as the richness of the concepts, the depth of hierarchies and the complexity of reasoning over the graph, since ontologies are developed using OWL languages and the complexity of these languages is known according to their expressivity. However, the developed ontology's complexity cannot simply be judged through the OWL language used since it provides the upper bound, therefore it is important to know which constructs are used.

Evaluation: The U Ontology is developed using the OWL DL syntax. The ontology makes use of RDFS elements which are a subset of the closure of OWL DLP.

This allows the ontology to be reasoned using a lightweight RDFS style (RDFS entailment rules) reasoner. As mentioned in [142], the use of RDFS elements allows RDFS style reasoners to compute practically relevant inferences with the knowledge base and other ontologies without making the ontology OWL Full. The metrics defined to ensure ontology complexity are:

- Number of subsumptions $= 10$
- Number of transitives $= 0$
- Number of nominals $= 0$
- Number of unions $= 0$

Based on the above discussion and the metrics value, the U Ontology remains within the OWL-DL expressivity, which makes it an ideal ontology for the Web because its use in knowledge bases and other ontologies would not make it OWL-Full.

Conclusion: `Verified`

9.5.2   Method 13: Searching for Anti-patterns

This method checks the presence of certain patterns in the ontology.

Method 13 (Searching for Anti-Patterns) [280]
"SPARQL queries over the ontology graph can be used to discover potentially problematic patterns. For example, results to the following queries have almost always been found to be problematic.
Detecting the anti-pattern of subsuming nothing:

```
select ?a where {
  ?a rdfs:subClassOf owl:Nothing .
}
```

Detecting the anti-pattern of skewed partitions:

```
select distinct ?A ?B1 ?B2 ?C1 where {
  ?B1 rdfs:subClassOf ?A .
  ?B2 rdfs:subClassOf ?A .
  ?C1 rdfs:subClassOf ?B1 .
  ?C1 owl:disjointWith ?B2 .
}''
```

Similar to object-oriented language, there are ontology design patterns to formalize the common configuration of ontologies. Some of the patterns help to design more useful ontologies because they are based on tested and trusted practices, whereas several patterns should be avoided. These patterns cause ontologies to fail or create problems at later stages when usage increases. Ontology design patterns were proposed in [32, 221] and their implementation has been discussed in detail.

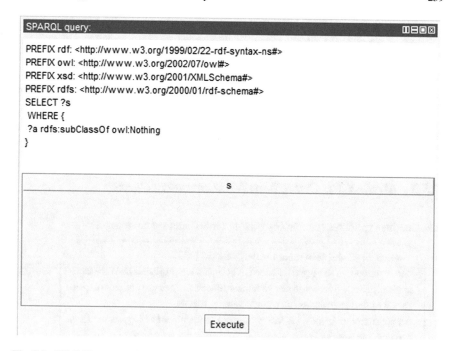

```
SPARQL query:                                                    ⊞⊟⊡⊠

PREFIX rdf: <http://www.w3.org/1999/02/22-rdf-syntax-ns#>
PREFIX owl: <http://www.w3.org/2002/07/owl#>
PREFIX xsd: <http://www.w3.org/2001/XMLSchema#>
PREFIX rdfs: <http://www.w3.org/2000/01/rdf-schema#>
SELECT ?s
 WHERE {
 ?a rdfs:subClassOf owl:Nothing
 }
```

s

                            Execute

Fig. 9.3 SPARQL query to identify an anti-pattern in the U ontology

Evaluation: Different patterns are checked by posing a SPARQL query to the U Ontology to verify the inclusion or exclusion of certain patterns. In addition to the queries presented in [280], we also used our own developed queries [8] are also used to identify the patterns which are not recommended in the ontologies and knowledge base.

While certain patterns should be used in an ontology because they guarantee good results, it is also important to consider several anti-patterns. The presence of anti-patterns signals the presence of a problem in the ontology. Figure 9.3 displays a SPARQL query to detect an anti-pattern of subsuming nothing in the U Ontology. The query did not find the presence of such a pattern in the ontology.

The other anti-patterns considered for detection in the ontology were:

```
select distinct ?A ?B1 ?B2 ?C1 where {
   ?B1   rdfs:subClassOf   ?A .
   ?B2   rdfs:subClassOf   ?A .
   ?C1   rdfs:subClassOf   ?B1.
   ?C1   owl:disjointWith ?B2.
}

Select distinct ?ind
where{
   ?a   owl:disjointWith   ?b.
```

```
    ?ind ref:type            ?a.
    ?ind rdf:type            ?b.
}
```

None of the above anti-patterns were found in the U Ontology, which assures that the ontology does not have any issues associated with these anti-patterns.

Conclusion: Verified

9.5.3   Method 14: OntoClean Meta-property Check

This method validates the ontology using OntoClean methodology.

> **Method 14 (OntoClean meta-property check) [280]**
> "An ontology can be tagged with the OntoClean meta-properties and then automatically checked for constraint violations. Since the tagging of classes is expensive, we provide an automatic tagging system, AEON.
> All constraint violations, i.e. inconsistencies in the meta-ontology, come from two possible sources:
> - an incorrect meta-property tagging, or
> - an incorrect subsumption.
> The evaluator has to carefully consider each inconsistency, determine which type of error has been discovered, and then either correct the tagging or redesign the subsumption hierarchy."

OntoClean [128] is an ontology evaluation and validation methodology. It measures the adequacy of the ontology by analysing the taxonomic relationships present in the ontology. It makes use of philosophical notions such as rigidity, unity, dependency, and identity (known as OntoClean meta-properties) to formally analyse the classes and their subsumption hierarchies. While OntoClean has been presented in conferences [127, 211], and is well-documented and well-acknowledged by the community, it is still used infrequently due to its complexity in applications and limited support from annotation tools [128]. There are a few tools in which OntoClean meta-properties are implemented, such as WebODE [96] and OntoEdit [252].

Evaluation: For the validation of the U Ontology, OntoCleans meta-properties are not used since it takes significant effort to annotate all the constructs using meta-properties. The support of OntoClean in Protg is minimal and the plug-in is not updated to the latest version of Protg (i.e version 4.2).

Conclusion: Not Applicable

9.6   U Ontology Evaluation: Semantics Aspect

The semantic aspect of an ontology measures the semantics of the ontology. In the previous aspect, the structure of the ontology was analysed by measuring the typological characteristics of the ontology without considering its semantics. To improve the overall quality of the ontology, it is important to measure the semantic aspects as well, since the essence of ontologies is to carry and communicate the semantics of the domain of interest. Considering semantics along with the structure of the ontology allows RDFS/OWL semantics to be taken into consideration, otherwise ontology metrics generally consider the RDF graph model.

9.6.1   Method 15: Ensuring a Stable Class Hierarchy

This method checks the ontology hierarchies (incorporating the semantic aspect) to determine whether or not they are stable.

> **Method 15 (Ensuring a stable class hierarchy) [280]**
> "Calculate a normalized class depth measure, i.e. calculate the length of the longest subsumption path on the normalized version of the ontology $md(N(O))$. Now calculate the stable minimal depth of the ontology $md^{min}(O)$. If
>
> $$md(N(O)) \neq md^{min}(O)$$
>
> then the ontology hierarchy is not stable and may collapse."

In order to incorporate the semantic in metrics, the explicit model (structure) of the ontology should not be considered; rather, all the models that are entailed from the ontology should be considered. This means that metrics need to be based on implicit statements and not on explicit statements, because the derived implicit statements represent the coverage of the domain knowledge conceptualized by the ontology. Therefore, reasoners are used to measure semantics and a normalization technique is used to obtain stable metrics [283].

Evaluation: To measure the stability of the class hierarchy in the U Ontology, a U Ontology Lite (UOT) is created to apply normalization steps. The UOT contains two classes: OntologyUsage and ConceptUsage and properties. In the first normalization step, the anonymous classes are removed; in the case of UOT, no such anonymous class was present. In the second step, anonymous individuals are removed. In UOT, no blank node was present and all the individuals were names with URI references. In the third step, subsumption hierarchies are materialized; to do this, RDFS entailment rules were used to generate the new statement materializing

the subsumption relationship. In the fourth step of normalization, all the concepts
and properties are instantiated. This means that the instances of classes and properties are populated. In the fifth step, which is similar to the third step, properties
are materialized. The normalized class depth and the stable minimal path of OUT
generated 2 and 2 respectively, but since these measures are obtained on the smaller
version of the ontology, they cannot be applied to the U Ontology. The measurement
of these metrics and evaluation of this method will be considered in future work.

Conclusion: Deferred

9.6.2 Method 16: Measuring Language Completeness

This method measures the language completeness of the ontology.

> **Method 16 (Measuring language completeness) [280]**
> "We define a function Υ_i with the index i being a language fragment (if none
> is given, the assertional fragment is assumed) from an ontology O to the set
> of all possible axioms over the signature of O given the language fragment
> i. We introduce C_i as language completeness over the language fragment i.
> $$Ci(O) = |\{X \,|X \in \Upsilon(O), O \models X \vee O \models \neg X|\, / \,|\Upsilon(O)|"$$

Given the set of names (URI references) in an ontology, language completeness
measures the ratio between the knowledge that can be expressed and the knowledge
that is stated.

Evaluation: When measuring language completeness, it is important to understand the semantic aspects of the ontology; however, since the U Ontology does not
have complex axioms (except a rdfs:subClassOf), an evaluation of this metric
will not offer usable insight. Therefore, this metric is not considered in the evaluation
of the U Ontology.

Conclusion: Not Applicable

9.7 U Ontology Evaluation: Representation Aspect

The representation aspect of an ontology analyses how the semantics of the ontology
are structurally represented. This aspect helps to identify mistakes which may arise
between the formal specification and the conceptualization. It is possible that the
semantics of the ontology are structurally represented in more than one way, thus
the need arises to obtain a normalized version to ensure that the sub-model of the

ontology has the same semantics. In [181, 283], the authors proposed metrics to measure the depth of the taxonomy and find the normalized model with the same semantics.

9.7.1 Method 17: Explicitness of the Subsumption Hierarchy

This method identifies the relationships between the semantics and the structure of the ontology.

Method 17 (Explicitness of the subsumption hierarchy) [280]
"Calculate $ET(O)$.
- If $ET(O) = 1$ everything seems fine
- If $ET(O) < 1$ then some of the classes in the ontology have collapsed. Find the collapsed classes and repair the explicit class hierarchy
- If $ET(O) > 1$ part of the class hierarchy has not been explicated. Find that part and repair the class hierarchy."

Using the ontology normalization [283] functions, the maximum subsumption path length is computed and compared with the depth of the taxonomy. Note that here the taxonomy represents the normalized sub-model of the ontology offering the same semantics of the original sub-model (prior normalization). Reference [280] computes two metrics: maximum depth of the taxonomy (TD) of ontology O and maximum subsumption path length (SL) of the normalized version of ontology O.

Evaluation: For this method $ET(O) = TD(O)/SL(O)$ is computed and the following measures are obtained

$$ET(U Ontology) = 3/3 = 1$$

According to the definition of the metric, if both *TD* and *SL* are the same, it can be safely assumed that with the present structural representations of the ontology, there seems to be a balance in the taxonomy hierarchy and the semantics (shared conceptualization).

Conclusion: Verified

9.7.2 Method 18: Explicit Terminology Ratio

This method identifies the explicit terminology ratio in the ontology.

> **Method 18 (Explicit terminology ratio) [280]**
>
> "Calculate $R_C(O)$ and $R_P(O)$.
> - If $R_C(O) = R_P(O) = 1$, this indicates there is no problem with the coverage of elements with names in the ontology
> - If $R_C(O) < 1$ or $R_P(O) < 1$ and the ontology does not include a mapping to an external vocabulary, this indicates there are possible problems since a number of names have collapsed to describe the same class
> - If $R_C(O) < 1$ or $R_P(O) < 1$ and the ontology includes a mapping to an external vocabulary, we can remove all axioms providing the mapping and calculate $R_C(O')$ and $R_P(O')$ anew
> - If $R_C(O) > 1$ or $R_P(O) > 1$, this indicates that not all interesting classes or properties have been given a name, i.e. the coverage of classes and properties with names may not be sufficient"

This method is based on the measure labelled M29 proposed by Gangemi et al. [105] called the Class/relations ratio which returns the ratio between classes and the relations in the ontology graph. For a given ontology, this means the number of nodes representing classes and the number of nodes representing relations within the ontology graph.

Evaluation: As mentioned in the previous method (method 17), the $ET(U$ $Ontology) = 1$ therefore the value of $R_C(O) = |C_N(O)|/|C(O)| = 1$ (where $C_N(O) = 30$, and $C(O) = 30$) and $R_P(O) = |P_N(O)|/|P(O)| = 1$ (where $P_N(O) = 45$, and $P(O) = 45$) the ratio between the normalized and not normalized ontology graphs remains the same.

Conclusion: `Verified`

9.8   U Ontology Evaluation: Context Aspect

The context aspect in ontology evaluation refers to the identification and creation of the relevant artefact accompanying an ontology. The identified and created additional artefacts are then used by the evaluating tool to support the validation and verification processes. These additional artefacts specify the context of the ontology. One of the early approaches in providing the context is the use of competency questions [270]. Competency questions allow evaluators to verify whether the developed ontology is able to answer all the issues raised in the competency question. To automate the verification process, competency questions need to be formally represented, which has still not been fully explored. Aside from this, certain constraints are imposed to verify the ontology. One of the latest approaches in this regard is the use of a unit test in ontology evaluation [280].

9.8.1 Method 19: Checking Competency Questions Against Results

This methods verifies the adequacy of an ontology using competency questions.

> **Method 19 (Checking competency questions against results) [280]**
> "Formalize the competency questions as a SPARQL query. Write down the expected answer as a SPARQL query result, either in XML or in JSON. Compare the actual and expected results. Note that the order of the results is often undefined."

Competency questions describe what kind of knowledge the resulting ontology is supposed to return [270]. The preferred approach to automate the verification process is to formalize these competency questions instead of merely having them written down in natural language.

Evaluation: Competency questions were presented in Sect. 7.4.2 to identify the scope and requirements for the ontology usage domain. The conceptual model of the U Ontology was developed based on these competency questions and was formalized using OWL language. The OUSAF was evaluated using the U Ontology in which SPARQL queries are generated, based on the competency questions presented in Sect. 7.4.2. The SPARQL queries are presented in Sects. 8.4–8.6 for identification and empirical and quantitative analysis, respectively.

Conclusion: `Verified`

9.8.2 Method 20: Checking Competency Questions with Constraints

This method validates the ontology by using competency questions with constraints.

> **Method 20 (Checking competency questions with constraints) [280]**
> "Formalize the competency questions for ontology O as a SPARQL CONSTRUCT query that formulates the result in RDF as an ontology R. Merge R with O and a possibly empty ontology containing further constraints C. Check the merged ontology for inconsistencies."

In the previous method, formalized competency questions were used in the form of SPARQL queries to verify the ontology. However, it is possible that when writing the competency questions, not all the requirements are captured or known. Therefore, to verify whether the ontology will be able to accommodate future changes, competency questions with constraints are used for evaluation.

Evaluation: This method is not considered for the evaluation of the U Ontology. The method is helpful for ontologies which are dynamic in nature and require frequent changes. Changes are expected in the U Ontology but not on a regular basis, therefore this method will be considered in future work.

Conclusion: Deferred

9.8.3   Method 21: Unit Testing with Test Ontologies

This method validates the ontology using test ontologies.

> **Method 21 (Unit testing with test ontologies) [280]**
> "For each axiom A_i^+ in the positive test ontology T^+ test whether the axiom is being inferred by the tested ontology O. For every axiom that is not being inferred, issue an error message. For each axiom A_i^- in the negative test ontology T^- test whether the axiom is being inferred by the tested ontology O. For every axiom that is being inferred, issue an error message."

The concept of using a unit test is quite new in ontologies, although it has been extensively used in software development and testing. In the case of ontologies, a test ontology is used to verify whether or not certain axioms can be derived from the ontology [281].

Evaluation: As pointed out by Vrandecic [280], test ontologies are meant to be created and grown during the maintenance of the ontology. Every time an error is encountered in the usage of the ontology, the error is formalized and added to the appropriate ontology, therefore this method is not applicable with the current state of the ontology.

Conclusion: Not Applicable

9.8.4   Method 22: Increasing Expressivity

This model checks the consistency in expressiveness.

> **Method 22 (Increasing expressivity) [280]**
> "An ontology O can be accompanied by a highly axiomatized version of the ontology, C. The merged ontology of $O \cup C$ has to be consistent, otherwise the inconsistencies point to errors in O."

Ontologies which are lightweight in their design (i.e. they do not use constructs which make ontology reasoning undecidable) are the preferred choice for information systems and are often recommended for Web usage requiring less time for inference.

Evaluation: The U Ontology is evaluated for inconsistencies and disjoint violations using a reasoner. The FaCT++ [263] reasoner and an RDFS-based reasoner implemented by the Virtuoso server [212] are used to verify all the axiomatic triples implemented in the U Ontology. During verification, no violation or inconsistency was reported. However, the reasoner found the presence of unsupported datatypes[2] which was addressed by using a supported datatype i.e. `dateTime`.

Conclusion: `Verified`

9.8.5   Method 23: Inconsistency Checks with Rules

This method checks the presence of inconsistencies in an ontology with the help of rules.

Method 23 (Inconsistency checks with rules) [280]
"Translate the ontology to be evaluated and possible constraint ontologies to a logic program. This translation does not have to be complete. Formalize further constraints as rules or integrity constraints.
Concatenate the translated ontologies and the further constraints or integrity constraints. Run the resulting program. If any integrity constraints are raised, the evaluated ontology contains errors."

The consistency checks in ontologies can be verified by making use of logical rules expressed using formalism, for example, SWRL [155] and RDFS entailment rules [157]. With the help of the expressivity of these languages, the context ontologies are not limited to OWL languages, therefore customized rules can be used for verification.

Evaluation: As mentioned earlier, the FaCT++ reasoner is used to validate the U Ontology and identify inconsistencies, if any.

Conclusion: `Verified`

9.9   Summary of U Ontology Evaluation

The U Ontology is evaluated from multiple aspects, as illustrated in the above sections, using the methods specified by Vrandecic [280]. Based on the results obtained, an overall evaluation for each method is concluded by using the metrics defined in Sect. 9.2.3. To quantify the U Ontology evaluation and provide conclusive remarks about the results, the summarized values of these predefined metrics are shown in Fig. 9.4.

As shown in Fig. 9.4, of the 23 methods, 18 methods were `Verified` to meet the expectations of the methodology proposed by Vrandecic [280]. Of the five methods

---

[2]ReasonerInternalException: Unsupported datatype "http://www.w3.org/2001/XMLSchema# dateTimeStamp".

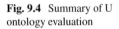

Fig. 9.4 Summary of U ontology evaluation

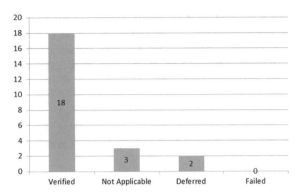

which were not evaluated as `Verified`, three were `Not Applicable` because of a lack of technical support in applying these methods. For example, Method 14 (OntoClean meta-property check) requires the use of meta-properties defined by OntoClean to annotate the ontology to measure the rigidity, unity, dependency and identity, while manual annotation with the OntoClean meta-property is impractical, and built-in automatic annotation support in the ontology tools is limited. Therefore, this method was marked as `Not Applicable`. Two of the five methods not verified are classified as `Deferred` which means that due to the complexity and extensive resource requirements needed to verify these methods, they will be considered in our future work. However, none of the methods failed in terms of not conforming to methodology expectations or not representing the required behaviour and characteristics.

Based on the above summarized evaluation of the U Ontology, it can conclusively be stated that the ontology represents the required quality standards and possesses the expected structural and semantic characteristics.

9.10 Recapitulation

The objective of this chapter was to evaluate the U Ontology to ensure that the developed ontology for representing Ontology Usage is of an acceptable quality. The evaluation methodology proposed by Vrandecic [280] is used for the U Ontology. This methodology comprises 23 methods (in accordance with the specification of the gold standard) which evaluate the ontology from six aspects. The U Ontology is analysed against each method and, based on the observations made, a metric is assigned. Each metric is then summarized and plotted on a chart to provide a summary of the U Ontology evaluation.

Chapter 10
New Research Directions

10.1 Introduction

As discussed in the previous chapters, numerous domain ontologies developed in different areas have resulted in a huge growth of structured data on the Web. While this has led to an increase in the number of ontologies published on the Web, at the same time there is *no formal approach to evaluate, measure, and analyse the use of those ontologies*. Such a study is very important to:

- Mine the hidden insights present in the use of formalized knowledge from the ontologies being used.
- Use these hidden insights to provide ontology developers with a usage feedback loop which they can use in the ontology maintenance process.

To realize these benefits, the OUSAF which provides pragmatic feedback and analysis of the use of domain ontologies on the Web is proposed in this book. The framework has been explained in detail and the working of each phase has been demonstrated in various scenarios to prove the effectiveness of the framework in providing the insights that are needed to manage the ontology lifecycle in an informed way.

However, during the course of the work presented in this book, several important future directions have emerged. Considering these possibilities in future work will strengthen the proposed methodology and framework and will help to integrate OUA with the ontology lifecycle model. Following are the high level areas for future work which have been identified. However, it should be noted that the identified list is not exhaustive and various other avenues of future work are possible.

10.2   Publish the Ontology Usage Analysis in the Form of an Ontology Usage Catalogue

To enable data publishers or other ontology users to access the latest state of an ontology's adoption and usage, the development of an ontology usage catalogue is proposed as future work. A catalogue provides users with the snapshot information which they need for their analysis at different levels of granularity. This requires to create an identifier for each ontology, present its entities and the relationships between them. It would be desirable to build an Ontology Usage Catalogue to represent the key entities of the domain in focus, and the usage level of the ontology components.

The ontology usage catalogue for each domain ontology would provide consolidated terminological knowledge, representing the key entities constituting the application domain. Additionally, it would provide quantitative measures for the semantic (RDF) repositories which could use it to evaluate the axioms which need to be supported for reasoning, based on the actual usage data. It is important that the developed ontology usage catalogue is regularly updated with usage information as they become available.

10.3   Explore Other Dimensions and Aspects Required for Measuring Ontology Usage and Provide Support for Reasoning over the Collected Dataset

In order to undertake a more specialized empirical and quantitative analysis of ontology usage, other aspects and dimensions need to be explored. It would be interesting to measure the use of RDFS and OWL standards in ontologies and instance data, and to examine how the ontologies and instance data are interlinked and mapped. The defined metrics in both EMP-AF and QUA-AF can be extended by incorporating the use of axiomatic triples. For example, when measuring the RelationshipValue (Chap. 5) of an object property, the concepts in the domain and range of property are calculated. However, it is possible that the object property has a sub-property axiom and the domain range value of sub-properties can be considered for measuring the richness value.

Similarly, the provision of subsumption axioms can be considered to explore the reasoning possible on the knowledge patterns reported by the usage analysis. It would also be interesting to know what kind of reasoning the defined semantics enable and how much can be obtained in the form of implicit knowledge from the explicit knowledge through reasoning.

10.4   Explore and Incorporate Other Approaches for Measuring Incentives

In Chap. 6, three dimensions are used to quantitatively analyse the use of ontologies. To incorporate the business dimension, the commercial incentives available to the particular ontologies are measured. Due to the lack of any formal study in the literature that quantifies the exact commercial benefits available to publishers in this book, an Incentive measure which measures the benefits to the ontology (in the form of search visibility) in search engines is used. However, this needs to be extended in future work to consider the other forms of incentives available to data publishers. These can be for example the financial incentives or level of recognition to the data publishers or ontology developers as a result of using an ontology.

Similarly, a survey of data publishers can be conducted to discover which factors motivate them to use ontologies on the Web. Based on the survey findings, an adaptive incentive model for measuring ontology usage could be developed.

References

1. Alani, H., Brewster, C., Shadbolt, N.: Ranking ontologies with AKTiveRank. In: Proceedings of the 5th International conference on the Semantic Web (ISWS). Lecture Notes in Computer Science, vol. 4273, pp. 1–15. Springer, Athens, Georgia (2006)
2. Albert, R., Barabási, A.L.: Statistical mechanics of complex networks. Rev. Mod. Phys. **74**, 47–97 (2002). https://doi.org/10.1103/RevModPhys.74.47
3. Almpanidis, G., Kotropoulos, C., Pitas, I.: Combining text and link analysis for focused crawling: an application for vertical search engines. Inf. Syst. **32**(6), 886–908 (2007)
4. Amardeilh, F.: OntoPop or how to annotate documents and populate ontologies from texts. In: Proceedings of the Workshop on Mastering the Gap: From Information Extraction to Semantic Representation (ESWC2006), Budva, Montenegro (2006)
5. Aoyama, M., et al.: New age of software development: How component-based software engineering changes the way of software development. In: 1998 International Workshop on CBSE. Citeseer (1998)
6. Ashburner, M., Ball, C., Blake, J., Botstein, D., Butler, H., Cherry, M., Davis, A., Dolinski, K., Dwight, S., Eppig, J.: Gene ontology: tool for the unification of biology. Nat. Genet. **25**(1), 25–29 (2000). http://publication.wilsonwong.me
7. Ashraf, J.: List of datasources included in grds2 dataset (google document) (2011). https://docs.google.com/spreadsheet/ccc?key=0AqjAK1TTtaSZdGpIMkVQUTRNenlrTGc tR2J1bkl6WEE. Accessed 14 Nov 2017
8. Ashraf, J., Cyganiak, R., O'Riain, S., Hadzic., M.: Open eBusiness ontology usage: investigating community implementation of good relations. In: Proceedings of Linked Data on the Web Workshop (LDOW) at WWW2011. CEUR Workshop Proceedings, vol. 813, pp. 1–11. CEUR-WS.org, Hyderabad, India (2011)
9. Asunción Gómez-Pérez, A., Fernández-López, M.: DE VINCENTE, A.: Towards a method to conceptualize domain ontologies. ECAI-96 Workshop on Ontological Engineering. ECAI-96 Workshop Proceedings, pp. 41–52. Budapest, Hungary (1996)
10. Atzeni, P., Mecca, G., Merialdo, P.: Semistructured und structured data in the web: going back and forth. SIGMOD Rec. **26**(4), 16–23 (1997). http://dblp.uni-trier.de/db/journals/sigmod/sigmod26.html#AtzeniMM97
11. Auer, S., Bizer, C., Kobilarov, G., Lehmann, J., Cyganiak, R., Ives, Z.: Dbpedia: A nucleus for a web of open data. The Semantic Web pp. 722–735 (2007)
12. Auer, S., Lehmann, J.: Creating knowledge out of interlinked data. Semant. Web J. **1**(1), 97–104 (2010)
13. Baker, T., Herman, I.: Semantic web case studies and use cases (2009). http://www.w3.org/2001/sw/sweo/public/UseCases/. Accessed 12 May 2009

© Springer International Publishing AG 2018

J. Ashraf et al., *Measuring and Analysing the Use of Ontologies*, Studies in Computational Intelligence 767, https://doi.org/10.1007/978-3-319-75681-3

14. Barabási, A., Albert, R.: Emergence of scaling in random networks. Science **286**(5439), 509–512 (1999)
15. Barabási, A.L., Albert, R., Jeong, H.: Scale-free characteristics of random networks: the topology of the world-wide web. Physica A: Stat. Mech. Appl. **281**(1), 69–77 (2000)
16. Batsakis, S., Petrakis, E., Milios, E.: Improving the performance of focused web crawlers. Data Knowl. Eng. **68**(10), 1001–1013 (2009)
17. Beckett, D.: RDF/XML Syntax Specification (Revised) (2004). http://www.w3.org/TR/rdf-syntax-grammar/
18. Benjamin, P.C., Menzel, C.P., Mayer, R.J., et. al, F.F.: IDEF5 Ontology description capture method report. Knowledge based systems, inc method report (1994)
19. Bergman, M.: Making connections real (web page) (2011). http://www.mkbergman.com/941/making-connections-real/. Accessed 14 Nov 2017
20. Berners-Lee, T.: Semantic web road map. Design Issues for the World Wide Web, pp. 1–10 (1998). http://www.w3.org/DesignIssues/Semantic.html
21. Berners-Lee, T.: Web architecture from 50,000 feet (1998). http://www.w3.org/DesignIssues/Architecture.html. Accessed 13 Dec 2017
22. Berners-Lee, T.: Linked data—design issue (2006). http://www.w3.org/DesignIssues/LinkedData.html
23. Berners-Lee, T., Fielding, R.T., Masinter, L.: Uniform Resource Identifiers (URI): Generic Syntax. Internet RFC 2396 (1998)
24. Berners-Lee, T., Fielding, R.T., Masinter, L.: Uniform resource identifier (uri): Generic syntax. Netw. Work. Group **66**(3986), 1–61 (2005)
25. Berners-Lee, T., Fischetti, M.: Weaving the Web: The Original Design and Ultimate Destiny of the World Wide Web by Its Inventor. Harper, San Francisco (1999)
26. Berrueta, D., Phipps, J.: Best Practice Recipes for Publishing RDF Vocabularies. W3C Working Group Note (2008)
27. Bicer, V., Laleci, G.B., Dogac, A., Kabak, Y.: Providing semantic interoperability in the healthcare domain through ontology mapping. Sigmod Rec. **34**(3) (2005)
28. Bishop, B., Kiryakov, A., Ognyanov, D., Peikov, I., Tashev, Z., Velkov, R.: FactForge: A fast track to the Web of data. Semant. Web **2**(2), 157–166 (2011). http://dblp.uni-trier.de/db/journals/semweb/semweb2.html#BishopKOPTV11a
29. Bizer, C., Cyganiak, R., Heath, T.: How to Publish Linked Data on the Web (2008). http://www4.wiwiss.fu-berlin.de/bizer/pub/LinkedDataTutorial/
30. Bizer, C., Heath, T., Berners-Lee, T.: Linked data–The story so far. Int. J. Semant. Web Inf. Syst. **5**(3), 1–22 (2009)
31. Bizer, C., Jentzsch, A., Cyganiak, R.: State of the Linked Open Data (LOD) Cloud. Technical Report 5 April 2011 (2011). http://www4.wiwiss.fu-berlin.de/lodcloud/state/
32. Blomqvist, E., Gangemi, A., Presutti, V.: Experiments on pattern-based ontology design. In: Proceedings of the Fifth International Conference on Knowledge Capture, K-CAP '09, pp. 41–48. ACM, New York, NY, USA (2009). https://doi.org/10.1145/1597735.1597743
33. Bock, J., Haase, P., Ji, Q., Volz, R.: Benchmarking OWL Reasoners. In: van Harmelen, F., Herzig, A., Hitzler, P., Lin, Z., Piskac, R., Qi, G. (eds.) Proceedings of the ARea2008 Workshop, vol. 350. CEUR Workshop Proceedings (2008). http://ceur-ws.org
34. Boehm, B.: A spiral model of software development and enhancement. Computer **21**(5), 61–72 (1988)
35. Boehm, B.W.: A spiral model of software development and enhancement. Softw. Eng. Project Manag., 128–142 (1987)
36. Bollacker, K.D., Cook, R.P., Tufts, P.: Freebase: A shared database of structured general human knowledge. In: AAAI, pp. 1962–1963. AAAI Press (2007). http://dblp.uni-trier.de/db/conf/aaai/aaai2007.html#BollackerCT07
37. Borgatti, S.: 2-Mode concepts in social network analysis. In: Meyers, R. (ed.) Encyclopedia of Complexity and Systems Science. Springer (2009)
38. Borgatti, S., Halgin, D.: Analyzing affiliation networks, pp. 417–433. The Sage Handbook of Social Network, Analysis (2011)

39. Bozsak, E., Ehrig, M., Handschuh, S., Hotho, A., Maedche, A., Motik, B., Oberle, D., Schmitz, C., Staab, S., Stojanovic, L., Stojanovic, N., Studer, R., Stumme, G., Sure, Y., Tane, J., Volz, R., Zacharias, V.: Kaon—towards a large scale semantic web. In: Bauknecht, K., Tjoa, A.M., Quirchmayr, G. (eds.) Proceedings of the Third International Conference on E-Commerce and Web Technologies EC-Web 2002. Lecture Notes in Computer Science, vol. 2455, pp. 304–313. Springer (2002)

40. Brank, J., Grobelnik, M., Mladenić, D.: A Survey of Ontology Evaluation Techniques. In: Proceedings of the Conference on Data Mining and Data Warehouses (SiKDD), pp. 166–170. Ljubljana, Slovenia (2005)

41. Brass, D.J.: Men's and women's networks: a study of interaction patterns and influence in an organization. Acad. Manag. J. **28**(2), 327–343 (1985)

42. Braun, S., Schmidt, A., Walter, A., Nagypal, G., Zacharias, V.: Ontology maturing: a collaborative web 2.0 approach to ontology engineering. In: Noy, N., Alani, H., Stumme, G., Mika, P., Sure, Y., Vrandecic, D. (eds.) Proceedings of the Workshop on Social and Collaborative Construction of Structured Knowledge (CKC 2007) at the 16th International World Wide Web Conference (WWW2007) Banff, Canada, May 8, 2007, CEUR Workshop Proceedings, vol. 273 (2007)

43. Breslin, J.G., Decker, S., Harth, A., Bojars, U.: SIOC: an approach to connect web-based communities. Int. J. Web Based Commun. **2**, 133–142 (2006)

44. Breslin, J.G., O'Sullivan, D., Passant, A., Vasiliu, L.: Semantic web computing in industry. Comput. Ind. **61**(8), 729–741 (2010). https://doi.org/10.1016/j.compind.2010.05.002, http://www.sciencedirect.com/science/article/pii/S0166361510000515

45. Brewster, C., Alani, H., Dasmahapatra, S., Wilks, Y.: Data driven ontology evaluation. In: International Conference on Language Resources and Evaluation. Lisbon, Portugal (2004)

46. Brickley, D., Guha, R.V.: Rdf vocabulary description language 1.0: Rdf schema. W3C Recommendation **10** (2004). http://www.w3.org/TR/rdf-schema/

47. Brickley, D., Miller, L.: Foaf vocabulary specification. Namespace Document 2 Sept 2004, FOAF Project (2004). http://xmlns.com/foaf/0.1/

48. Broder, A., Kumar, R., Maghoul, F., Raghavan, P., Rajagopalan, S., Stata, R., Tomkins, A., Wiener, J.: Graph structure in the web. Comput. Netw. **33**(1), 309–320 (2000)

49. Broekstra, J., Kampman, A., van Harmelen, F.: Sesame: a generic architecture for storing and querying rdf and rdf schema. In: Horrocks, I., Hendler, J. (eds.) The Semantic Web ISWC 2002. Lecture Notes in Computer Science, vol. 2342, pp. 54–68. Springer, Berlin/Heidelberg (2002)

50. Buitelaar, P., Eigner, T., Declerck, T.: OntoSelect: A dynamic ontology library with support for ontology selection. In: Proceedings of the Demo Session at the International Semantic Web Conference (ISWC) Demo Track. Hiroshima, Japan (2004)

51. Caire, P., van der Torre, L.: Convivial Ambient Technologies: requirements. Ontology and design. Comput. J. **53**(8), 1229–1256 (2010)

52. Carrol, J., McBride, B.: The Jena Semantic Web Toolkit. Public api, HP-Labs, Bristol (2001). http://www.hpl.hp.com/semweb/jena-top.html

53. Carroll, J.J., Bizer, C., Hayes, P., Stickler, P.: Named graphs, provenance and trust. In: Proceedings of the 14th International Conference on World Wide Web, WWW '05, pp. 613–622. ACM, New York, NY, USA (2005)

54. Celjuska, D., Vargas-vera, D.M.: Ontosophie: A semi-automatic system for ontology population from text. In: International Conference on Natural Language Processing (ICON) (2004)

55. Chakrabarti, S., Van den Berg, M., Dom, B.: Focused crawling: a new approach to topic-specific web resource discovery. Comput. Netw. **31**(11), 1623–1640 (1999)

56. Changrui, Y., Yan, L.: Comparative research on methodologies for domain ontology development. In: D.S. Huang, Y. Gan, P. Gupta, M. Gromiha (eds.) Advanced Intelligent Computing Theories and Applications. With Aspects of Artificial Intelligence. Lecture Notes in Computer Science, vol. 6839, pp. 349–356. Springer, Berlin/Heidelberg (2012). https://doi.org/10.1007/978-3-642-25944-9_45

57. Chaudhri, V.K., Farquhar, A., Fikes, R., Karp, P.D., Rice, J.: Okbc: A programmatic foundation for knowledge base interoperability. In: Mostow, J., Rich, C. (eds.) AAAI/IAAI, pp. 600–607. AAAI Press/The MIT Press (1998). http://dblp.uni-trier.de/db/conf/aaai/aaai98.html#ChaudhriFFKR98

58. Cheng, G., Gong, S., Qu, Y.: An empirical study of vocabulary relatedness and its application to recommender systems. International Semantic Web Conference (ISWC). Lecture Notes in Computer Science, vol. 7031, pp. 98–113. Springer, Bonn, Germany (2011)

59. Cheng, G., Qu, Y.: Term dependence on the semantic web. In: Proceedings of 7th International Semantic Web Conference (ISWC). Lecture Notes in Computer Science, vol. 5318, pp. 665–680. Springer, Karlsruhe, Germany (2008)

60. Claudio, M., Nicola, G., Alessandro, O., Luc, S.: The wonderweb library of foundational ontologies. Wonderweb deliverable d18 (2005). http://wonderweb.semanticweb.org/deliverables/documents/D18.pdf

61. Clauset, A., Shalizi, C.R., Newman, M.E.J.: Power-law distributions in empirical data. SIAM Rev. 51(4), 661–703 (2009)

62. Cooley, R., Mobasher, B., Srivastava, J.: Web mining: Information and pattern discovery on the world wide web. In: ICTAI '97: Proceedings of the 9th International Conference on Tools with Artificial Intelligence, p. 558. IEEE Computer Society, Washington, DC, USA (1997). http://doi.ieeecomputersociety.org/10.1109/TAI.1997.632303

63. Corcho, O., Fernández, M., Gómez-Pérez, A., López-Cima, A.: Building legal ontologies with METHONTOLOGY and WebODE. In: Benjamins, R., Casanovas, P., Breuker, J., Gangemi, A. (eds.) Law and the Semantic Web, LNAI, pp. 142–157. Springer, Heidelberg, DE (2005)

64. Corcho, O., Fernandez-Lopez, M., Gomez-Perez, A.: Ontological engineering: What are ontologies and how can we build them? In: Cardoso, J. (ed.) Semantic Web Services: Theory, Tools and Applications, chap. 03, pp. 44–70. IGI Global (2007)

65. Cranefield, S., Purvis, M.K.: Uml as an ontology modelling language. In: Intelligent Information Integration. CEUR Workshop Proceedings, vol. 23 (1999). https://ourarchive.otago.ac.nz/handle/10523/932

66. d'Aquin, M., Baldassarre, C., Gridinoc, L., Angeletou, S., Sabou, M., Motta, E.: Characterizing knowledge on the semantic web with Watson. In: Proceedings of the 5th International Workshop on Evaluation of Ontologies and Ontology-based Tools (EON) at ISWC2007, CEUR Workshop Proceedings, vol. 329, pp. 1–10. CEUR-WS.org, Busan, Korea (2007)

67. d'Aquin, M., Lewen, H.: Cupboard—A place to expose your ontologies to applications and the community. In: Proceedings of the 6th European Semantic Web Conference (ESWC) on The Semantic Web: Research and Applications. Lecture Notes in Computer Science, vol. 5554, pp. 913–918. Springer Berlin/Heidelberg, Heraklion, Crete, Greece (2009)

68. d'Aquin, M., Motta, E.: Watson, more than a Semantic Web search engine. Semant. Web 2(1), 55–63 (2011)

69. d'Aquin, M., Noy, N.F.: Where to Publish and Find Ontologies? A Survey of Ontology Libraries. Web Semantics: Science, Services and Agents on the World Wide Web 11, 96–111 (2012). https://doi.org/10.1016/j.websem.2011.08.005, http://www.sciencedirect.com/science/article/pii/S157082681100076X

70. Dasgupta, S., Dinakarpandian, D., Lee, Y.: A Panoramic Approach to Integrated Evaluation of Ontologies in the Semantic Web. In: Proceedings of the 5th International Workshop on Evaluation of Ontologies and Ontology-based Tools(EON) at ISWC2007. CEUR Workshop Proceedings, vol. 329, pp. 31–40. CEUR-WS.org, Busan, Korea (2007)

71. David, J., Euzenat, J.: Comparison between Ontology Distances (Preliminary Results). In: A. Sheth, S. Staab, M. Dean, M. Paolucci, D. Maynard, T. Finin, K. Thirunarayan (eds.) The Semantic Web - ISWC 2008, *Lecture Notes in Computer Science*, vol. 5318, pp. 245–260. Springer, Berlin/Heidelberg (2008). http://dx.doi.org/10.1007/978-3-540-88564-1_16

72. Davies, S., Hatfield, J., Donaher, C., Zeitz, J.: User interface design considerations for Linked Data authoring environments. In: Proceedings of Linked Data on the Web Workshop (LDOW) at WWW2010, CEUR Workshop Proceedings, vol. 628. CEUR-WS.org, Raleigh, USA (2010)

73. De Bruijn, J., Lara, R., Polleres, A., Fensel, D.: OWL DL vs. OWL flight: conceptual modeling and reasoning for the semantic Web. In: Proceedings of the 14th International Conference on World Wide Web, pp. 623–632. ACM (2005)

74. Dean, M., Connolly, D., van Harmelen, F., Hendler, J., Horrocks, I., McGuinness, D.L., Patel-Schneider, P.F., Stein, L.A.: OWL:Web Ontology Language 1.0 reference. Technical report, W3C Working Draft. Technical Note (2002)

75. Decker, S., Erdmann, M., Fensel, D., Studer, R.: Ontobroker: Ontology based access to distributed and semi-structured information. In: Proceedings of the IFIP TC2/WG2. 6 Eighth Working Conference on Database Semantics-Semantic Issues in Multimedia Systems, pp. 351–369. Kluwer, BV (1999)

76. Della Valle, E., Cerizza, D., Bicer, V., Kabak, Y., Laleci, G., Lausen, H., DERI-Innsbruck, S.M.S.M.: The need for semantic web service in the ehealth. In: W3C Workshop on Frameworks for Semantics in Web Services (2005)

77. Ding, L., DiFranzo, D., Graves, A., Michaelis, J., Li, X., McGuinness, D.L., Hendler, J.: Data-gov wiki: Towards linking government data. In: AAAI Spring Symposium: Linked Data Meets Artificial Intelligence. AAAI (2010). http://dblp.uni-trier.de/db/conf/aaaiss/aaaiss2010-7.html#DingDGMLMH10

78. Ding, L., Finin, T.: Characterizing the semantic web on the web. international semantic web conference. Lecture Notes in Computer Science, vol. 4273, pp. 242–257. Springer, Athens, GA, USA (2006)

79. Ding, L., Finin, T., Joshi, A., Pan, R., Cost, R.S., Peng, Y., Reddivari, P., Doshi, V., Sachs, J.: Swoogle: a Search and Metadata Engine for the Semantic Web. In: Proceedings of the 13th ACM International Conference on Information and Knowledge Management, pp. 652–659. ACM, New York, NY, USA (2004)

80. Ding, L., Finin, T., Shinavier, J., McGuinness, D.L.: owl: same as and linked data: an empirical study (2010). http://journal.webscience.org/403/2/websci10_submission_123.pdf

81. Ding, L., Zhou, L., Finin, T., Joshi, A.: How the semantic web is being used: an analysis of FOAF documents. In: Proceedings of the 38th Annual Hawaii International Conference on System Sciences, vol. 4, pp. 113–120. IEEE Computer Society, Washington, DC, USA (2005)

82. Ding, Y.: A simple picture of Web evolution (2007). http://www.zdnet.com/blog/web2explorer/a-simple-picture-of-web-evolution/408. Accessed 12 Dec 2017

83. Ding, Y., Fensel, D.: Ontology library systems the key to successful ontology re-use. In: Proceedings of the 1st International Semantic Web Working Symposium (SWWS 2001), pp. 93–112 (2001)

84. Dodds, L.: Slug: a semantic web crawler. In: Proceedings of Jena User Conference, vol. 2006 (2006)

85. Dodds, L., Davis, I.: Linked Data Patterns (2010). http://patterns.dataincubator.org/book/

86. Dong, H., Hussain, F., Chang, E.: A survey in semantic web technologies-inspired focused crawlers. In: Third International Conference on Digital Information Management, 2008. ICDIM 2008, pp. 934 –936 (2008). https://doi.org/10.1109/ICDIM.2008.4746736

87. Ehrig, M., Maedche, A.: Ontology-focused crawling of web documents. In: Proceedings of the 2003 ACM Symposium on Applied Computing, pp. 1174–1178. ACM (2003)

88. Elkan, C., Greiner, R., Lenat, D.B., Guha, R.V.: building large knowledge-based systems: Representation and inference in the cyc project. Artif. Intell. **61**(1), 41–52 (1993). http://dblp.uni-trier.de/db/journals/ai/ai61.html#ElkanG93

89. Ell, B., Vrandecic, D., Simperl, E.P.B.: Labels in the Web of Data. In: Aroyo, L., Welty, C., Alani, H., Taylor, J., Bernstein, A., Kagal, L., Noy, N.F., Blomqvist, E. (eds.) International Semantic Web Conference (1). Lecture Notes in Computer Science, vol. 7031, pp. 162–176. Springer (2011)

90. Erétéo, G.: Semantic social network analysis. Ph.D. thesis, Orange Labs Telecom ParisTech INRIA Sophia Antipolis Méditerranée Karlsruhe Institute of Technology (2011). http://www-sop.inria.fr/members/Guillaume.Ereteo/PhD_thesis_Semantic_Social_Network_Analysis.pdf. Accessed 11 Dec 2017

91. Euzenat, J.: Building consensual knowledge bases: Context and architecture. In: Mars, N. (ed.) Towards Very Large Knowledge Bases - Proceedings of the KB&KS '95 Conference, pp. 143–155. IOS Press (1995)

92. Euzenat, J.: Corporate memory through cooperative creation of knowledge bases and hyper-documents (1996). http://citeseerx.ist.psu.edu/viewdoc/summary?doi=?doi=10.1.1.31.664

93. Euzenat, J.: A protocol for building consensual and consistent repositories. Technical Report, RR-3260, INRIA (1997). http://hal.inria.fr/inria-00073429

94. Fellbaum, C.: WordNet: An Electronic Lexical Database. MIT Press, Cambridge, MA (1998)

95. Fensel, D., Ding, Y., Omelayenko, B., Schulten, E., Botquin, G., Brown, M., Flett, A.: Product data integration in b 2 b e-commerce. IEEE Intell. Syst. **16**(4), 54–59 (2001)

96. Fernández-López, M., Gómez-Pérez, A.: The integration of ontoclean in webode. In: CEUR Workshop Proceedings (2002)

97. Fernandez-Lopez, M., Gomez-Perez, A., Euzenat, J., Gangemi, A., Kalfoglou, Y., Pisanelli, D., Schorlemmer, M., Steve, G., Stojanovic, L., Stumme, G., Sure, Y.: A survey on method-ologies for developing, maintaining, integrating, evaluating and reengineering ontologies. OntoWeb deliverable 1.4, Universidad Politecnia de Madrid (2002). http://www.aifb.uni-karlsruhe.de/WBS/ysu/publications/OntoWeb_Del_1-4.pdf

98. Fernández-López, M., Gómez-Pérez, A., Juristo, N.: Methontology: from ontological art towards ontological engineering. In: Proceedings of the AAAI97 Spring Symposium Series on Ontological Engineering (1997)

99. Fershtman, M.: Cohesive group detection in a social network by the segregation matrix index. Soc. Netw. **19**(3), 193–207 (1997)

100. Fox, M., Gruninger, M.: Enterprise modeling. AI Magazine **19**(3), 109 (1998)

101. Franceschet, M.: Collaboration in computer science: a network science approach. J. Am. Soc. Inf. Sci. Technol. **62**(10), 1992–2012 (2011). https://doi.org/10.1002/asi.21614

102. Freeman, L.C.: Finding social groups: A meta-analysis of the southern women data (2003)

103. Fürber, C., Hepp, M.: Towards a vocabulary for data quality management in semantic web architectures. In: Proceedings of the 1st International Workshop on Linked Web Data Man-agement, pp. 1–8. ACM (2011)

104. Gangemi, A., Catenacci, C., Ciaramita, M., Lehmann, J.: A Theoretical framework for on-tology evaluation and validation. In: Proceedings of 2nd Italian Semantic Web Workshop Semantic Web Application and Perspectives(SWAP), CEUR Workshop Proceedings. Trento, Italy (2005)

105. Gangemi, A., Catenacci, C., Ciaramita, M., Lehmann, J.: Ontology evaluation and vali-dation: an integrated formal model for the quality diagnostic task. Technical report, Lab-oratory of Applied Ontologies – CNR, Rome, Italy (2005). http://www.loa-cnr.it/Files/OntoEval4OntoDev_Final.pdf

106. Gangemi, A., Pisanelli, D.M., Steve, G.: An overview of the onions project: Applying ontologies to the integration of medical terminologies. Data Knowl. Eng. **31**(2), 183–220 (1999). https://doi.org/10.1016/S0169-023X(99)00023-3, http://www.sciencedirect.com/science/article/pii/S0169023X99000233

107. Gangemi, A., Pisanelli, D.M., Steve, G.: An overview of the onions project: applying ontolo-gies to the integration of medical terminologies. Data Knowl. Eng. **31**(2), 183–220 (1999)

108. Gangemi, A., Steve, G., Giacomelli, F.: Onions: An ontological methodology for taxonomic knowledge integration (1996). http://citeseerx.ist.psu.edu/viewdoc/summary?doi=?doi=10.1.1.22.3972

109. Garcia, A., ONeill, K., Garcia, L.J., Lord, P., Stevens, R., Corcho, O., Gibson, F.: Developing ontologies within decentralised settings. In: Chen, H., Wang, Y., Cheung, K.H., Sharda, R., Vo, S. (eds.) Semantic e-Science. Annals of Information Systems, vol. 11, pp. 99–139. Springer, US (2010). http://dx.doi.org/10.1007/978-1-4419-5908-9_4

110. Garcia, R., Gil, R.: Publishing XBRL as linked open data. In: Bizer, C., Heath, T., Berners-Lee, T., Hausenblas, M. (eds.) Proceedings of the Linked Data on the Web Workshop (LDOW) at WWW2009. CEUR Workshop Proceedings, vol. 538. CEUR-WS.org, Madrid, Spain (2009)

111. Garton, L., Haythornthwaite, C., Wellman, B.: Studying online social networks. J. Comput.-Mediat. Commun. **3**(1), (1997). https://doi.org/10.1111/j.1083-6101.1997.tb00062.x
112. Geleijnse, G., Korst, J.: Automatic ontology population by Googling. In: Proceedings of the Seventeenth Belgium-Netherlands Conference on Artificial Intelligence (BNAIC 2005), pp. 120–126. Brussels, Belgium (2005)
113. Goh, C.H., Bressan, S., Madnick, S., Siegel, M., Hian, C., Stephane, G., Stuart, B., Siegel, M.M.: Context interchange: new features and formalisms for the intelligent integration of information. ACM Trans. Inf. Syst. **17**, 270–293 (1999)
114. Gomez-Perez, A., Corcho, O.: Ontology languages for the semantic web. IEEE Intell. Syst. **17**(1), 54–60 (2002). https://doi.org/10.1109/5254.988453
115. Gómez-Pérez, A., Fernández-López, M., Corcho, O.: Ontological Engineering. Springer, Heidelberg, Berlin (2004)
116. Gomez-Perez, A., Suarez-Figueroa, M.C.: Scenarios for building ontology networks within the neon methodology. In: Proceedings of the Fifth International Conference on Knowledge Capture (K-CAP 2009)
117. González, R.G.: A semantic web approach to digital rights management. Ph.D. thesis, Universitat Pompeu Fabra, Barcelona, Spanien (2005)
118. Gradmann, S.: rdfs:frbr - Towards an implementation model for library catalogs using semantic web technology. Catal. Classif. Q. **39**(3–4), 63–75 (2005)
119. Graham, I., O'Callaghan, A., Wills, A.: Object-Oriented Methods: Principles & Practice, vol. 6. Addison-Wesley (2001)
120. Grant, J., Beckett, D., McBride, B.: Rdf Test Cases: N-triples. World Wide Web Consortium (W3C) working draf (2002). http://www.w3.org/TR/rdftestcases/#ntriples
121. Grover, V., Kettinger, W.: Process think: winning perspectives for business change in the information age. IGI Global (2000). https://doi.org/10.4018/978-1-87828-968-1
122. Gruber, T.R.: A Translation approach to portable ontology specifications. Knowl. Acquisit. **5**(2), 199–220 (1993). http://tomgruber.org/writing/ontolingua-kaj-1993.pdf
123. Gruninger, M., Fox, M.S.: The design and evaluation of ontologies for enterprise engineering. In: Workshop on Implemented Ontologies, European Conference on Artificial Intelligence (1994)
124. Gruninger, M., Fox, M.S.: The role of competency questions in enterprise engineering. In: Proceedings of the IFIP WG5.7 Workshop on Benchmarking - Theory and Practice (1994)
125. Grüninger, M., Fox, M.S.: Methodology for the design and evaluation of ontologies. In: International Joint Conference on Artificial Inteligence (IJCAI95), Workshop on Basic Ontological Issues in Knowledge Sharing (1995)
126. Guarino, N.: Formal Ontology and Information Systems, pp. 3–15. IOS Press (1998)
127. Guarino, N., Welty, C.: Evaluating ontological decisions with ontoclean. Commun. ACM **45**(2), 61–65 (2002)
128. Guarino, N., Welty, C.: An overview of OntoClean. In: Handbook on Ontologies, pp. 151–159. Springer, Germany (2004)
129. Guebitz, B., Schnedl, H., Khinast, J.: A risk management ontology for quality-by-design based on a new development approach according gamp 5.0. Expert Systems with Applications (2012)
130. Guéret, C., Groth, P.T., van Harmelen, F., Schlobach, S.: Finding the achilles heel of the web of data: using network analysis for link-recommendation. In: Patel-Schneider, P.F., Pan, Y., Hitzler, P., Mika, P., Pan, J.Z., Horrocks, I., Glimm, B. (eds.) International Semantic Web Conference (ISWC)). Lecture Notes in Computer Science, vol. 6496, pp. 289–304. Springer (2010)
131. Guizzardi, G.: On ontology, ontologies, conceptualizations, modeling languages, and (meta)models. In: Vasilecas, O., Eder, J., Caplinskas, A. (eds.) DB&IS, Frontiers in Artificial Intelligence and Applications, vol. 155, pp. 18–39. IOS Press (2006)
132. Haase, P., Stojanovic, L.: Consistent evolution of OWL ontologies. In: Gomez-Perez, A., Euzenat, J. (eds.) Proceedings of the Second European Semantic Web Conference, vol. 3532, pp. 182–197. Springer, Heraklion, Crete, Greece (2005)

133. Haase, P., Volz, R., Erdmann, M., Studer, R.: Ontology engineering and plugin development with the neon toolkit. In: ISWC/ASWC 2007 tutorial (2007)

134. Happel, H.J., Seedorf, S.: Applications of ontologies in software engineering. In: Proceedings of Workshop on Sematic Web Enabled Software Engineering (SWESE) on the ISWC, pp. 5–9. Citeseer (2006)

135. Harary, F.: Graph Theory. Addison-Wesley (1991)

136. Harth, A., Umbrich, J., Decker, S.: Multicrawler: A pipelined architecture for crawling and indexing semantic web data. The Semantic Web-ISWC **2006**, 258–271 (2006)

137. Hartmann, J., Sure, Y., Haase, P., Palma, R., del Carmen Surez-Figueroa, M.: Omv—ontology metadata vocabulary. In: Welty, C. (ed.) ISWC 2005—In Ontology Patterns for the Semantic Web (2005)

138. Hausenblas, M., Halb, W., Raimond, Y., Heath, T.: What is the size of the semantic web? In: In Proceedings of the International Conference on Semantic Systems (ISemantics), Universal Computer Science, pp. 6–16. Graz, Austria (2008)

139. Hayes, P.: RDF Semantics. Technical report (2004). http://www.w3.org/TR/rdf-mt/

140. Heath, T., Bizer, C.: Linked Data: Evolving the Web into a Global Data Space. Synthesis Lectures on the Semantic Web. Morgan & Claypool (2011). http://dx.doi.org/10.2200/S00334ED1V01Y201102WBE001

141. Hepp, M.: Possible ontologies: how reality constrains the development of relevant ontologies. IEEE Internet Comput. **11**(1), 90–96 (2007)

142. Hepp, M.: GoodRelations: An ontology for describing products and services offers on the web. In: Proceedings of the 16th International Conference on Knowledge Engineering: Practice and Patterns (EKAW). Lecture Notes in Computer Science, vol. 5268, pp. 329–346. Springer Berlin/Heidelberg, Sicily, Italy (2008)

143. Herman, I.: Ldow2011 workshop (blog post) (2011). https://ivanherman.wordpress.com/2011/03/29/ldow2011-workshop/. Accessed 11 Dec 2017

144. Heymann, P., Garcia-Molina, H.: Collaborative creation of communal hierarchical taxonomies in social tagging systems. Technical Report 2006-10, Stanford InfoLab (2006). http://ilpubs.stanford.edu:8090/775/

145. Hitzler, P., van Harmelen, F.: A reasonable semantic web. Semantic Web J. **1**(1–2), 39–44 (2010)

146. Hogan, A.: Exploiting rdfs and owl for integrating heterogeneous, large-scale, linked data corpora. Ph.D. thesis, National University of Ireland, Galway, Ireland (2011)

147. Hogan, A.: Owl for integrating heterogeneous, large-scale, linked data corpora. Ph.D. thesis, Ph. D. thesis, Digital Enterprise Research Institute, National University of Ireland, Galway (2011). http://aidanhogan.com/docs/thesis

148. Hogan, A., Harth, A., Passant, A., Decker, S., Polleres, A.: Weaving the pedantic web. In: Proceedings of Linked Data on the Web Workshop (LDOW) at WWW2010, CEUR Workshop Proceedings, vol. 628. CEUR-WS.org, Raleigh, USA (2010)

149. Hogan, A., Harth, A., Umbrich, J., Kinsella, S., Polleres, A., Decker, S.: Searching and browsing linked data with SWSE: the semantic web search engine. Web Semant.: Sci. Serv. Agents. World Wide Web **9**(4), 365–401 (2011)

150. Hogan, A., Zimmermann, A., Umbrich, J., Polleres, A., Decker, S.: Scalable and distributed methods for entity matching, consolidation and disambiguation over linked data corpora. J. Web Semant. **10**, 76–110 (2012). http://dblp.uni-trier.de/db/journals/ws/ws10.html#HoganZUPD12

151. Hogg, T., Wilkinson, D.M., Szabó, G., Brzozowski, M.J.: Multiple relationship types in online communities and social networks. In: AAAI Spring Symposium: Social Information Processing, pp. 30–35. AAAI (2008). http://dblp.uni-trier.de/db/conf/aaaiss/aaaiss2008-6.html#HoggWSB08

152. Hori, M., Euzenat, J., Patel-Schneider, P.F.: OWL Web Ontology Language XML presentation syntax (2003). W3C Note 11 June 2003

153. Horridge, M., Drummond, N., Goodwin, J., Rector, A., Stevens, R., Wang, H.H.: The manchester owl syntax. In: Grau, B.C., Hitzler, P., Shankey, C., Wallace, E. (eds.) Proceedings

of OWL: Experiences and Directions (OWLED'06). Athens, Georgia, USA, (2006). http://sunsite.informatik.rwth-aachen.de/Publications/CEUR-WS//Vol-216/

154. Horrocks, I.: Ist project 2001-33052 wonderweb: Ontology infrastructure for the semantic web. d29: Final report. Tech. rep. (2005). http://wonderweb.semanticweb.org/deliverables/documents/D29.pdf

155. Horrocks, I., Patel-Schneider, P., Boley, H., Tabet, S., Grosof, B., Dean, M., et al.: SWRL: A semantic web rule language combining OWL and RuleML. W3C Member Submission **21**, 79 (2004)

156. Horrocks, I., Patel-Schneider, P.F., van Harmelen, F.: From shiq and rdf to owl: The making of a web ontology language. J. Web Semant. **1**(1) (2003)

157. ter Horst, H.J.: Completeness, decidability and complexity of entailment for rdf schema and a semantic extension involving the owl vocabulary. Web Semant.: Sci. Serv. Agents World Wide Web **3**(2–3), 79 – 115 (2005). Selcted Papers from the International Semantic Web Conference, 2004 - ISWC, 2004

158. Hoser, B., Hotho, A., Jäschke, R., Schmitz, C., Stumme, G.: Semantic network analysis of ontologies. In: Sure, Y., Domingue, J. (eds.) The Semantic Web: Research and Applications. Lecture Notes in Computer Science, vol. 4011, pp. 514–529. Springer, Berlin/Heidelberg (2006)

159. Iqbal, A., Ureche, O., Hausenblas, M., Tummarello, G.: Ld2sd: Linked data driven software development. In: SEKE, pp. 240–245. Knowledge Systems Institute Graduate School (2009)

160. Isele, R., Umbrich, J., Bizer, C., Harth, A.: LDspider: An open-source crawling framework for the web of linked data. In: Proceedings of the International Semantic Web Conference (ISWC) Posters & Demonstrations Track, CEUR Workshop Proceedings, vol. 658, pp. 29–32. CEUR-WS.org, Shanghai, China (2010)

161. Jain, P., Hitzler, P., Yeh, P., Verma, K., Sheth, A.: Linked Data is Merely More Data. In: Proceedings of the AAAI Spring Symposium, Linked AI: Linked Data Meets Artificial Intelligence, pp. 82–86. AAAI, Menlo Park, CA, USA (2010)

162. Jarrar, M., Meersman, R.: Formal ontology engineering in the dogma approach. In: Meersman, R., Tari, Z. (eds.) CoopIS/DOA/ODBASE. Lecture Notes in Computer Science, vol. 2519, pp. 1238–1254. Springer (2002). http://dblp.uni-trier.de/db/conf/coopis/coopis2002.html#JarrarM02

163. Kloppenborg, T.: Contemporary project management. Cengage Learning (2011)

164. Klyne, G., Carroll, J.J.: Resource description framework (rdf): Concepts and abstract syntax. World Wide Web Consortium, Recommendation REC-rdf-concepts-20040210 (2004). http://www.w3.org/TR/rdf-concepts/

165. Knight, K., Chander, I., Haines, M., Hatzivassiloglou, V., Hovy, E.H., Iida, M., Luk, S.K., Whitney, R., Yamada, K.: Filling knowledge gaps in a broad-coverage machine translation system. CoRR **abs/cmp-lg/9506009** (1995). http://dblp.uni-trier.de/db/journals/corr/corr9506.html#abs-cmp-lg-9506009

166. Knight, K., Luk, S.K.: Building a large-scale knowledge base for machine translation. In: National Conference on Aritificial Intelligence, AAAI, pp. 773–778 (1994). http://arxiv.org/PS_cache/cmp-lg/pdf/9407/9407029.pdf

167. Knoke, D., Yang, S., Kuklinski, J.: Social Network Analysis, vol. 2. Sage Publications, Los Angeles, CA (2008)

168. Knublauch, H., Fergerson, R., Noy, N., Musen, M.: The protégé owl plugin: an open development environment for semantic web applications. Semant. Web-ISWC **2004**, 229–243 (2004)

169. Kobilarov, G., Bizer, C., Auer, S., Lehmann, J.: Dbpedia—a linked data hub and data source for web applications and enterprises. In: Proceedings of Developers Track of 18th International World Wide Web Conference (WWW 2009), April 20th–24th, Madrid, Spain (2009). http://www2009.eprints.org/228/

170. Kobilarov, G., Scott, T., Raimond, Y., Oliver, S., Sizemore, C., Smethurst, M., Bizer, C., Lee, R.: Media meets Semantic Web–how the BBC uses DBpedia and linked data to make connections. Semant. Web: Res. Appl., 723–737 (2009)

171. Kolias, V.D., Stoitsis, J., Golemati, S., Nikita, K.S.: Utilizing semantic web technologies in healthcare. In: Concepts and Trends in Healthcare Information Systems, pp. 9–19. Springer (2014)

172. Kotis, K.: On supporting hcome-3o ontology argumentation using semantic wiki technology. In: K. Kotis (ed.) OnTheMove (OTM) 2008 Conference, Community-Based Evolution of Knowledge-Intensive Systems (COMBEK'08) Workshop. LNCS, vol. 5333, pp. 193–199. Springer, Berlin Heidelberg, Mexico (2008). http://www.icsd.aegean.gr/kotis/publications/combek2008.pdf

173. Kotis, K., Vouros, G.: Human-centered ontology engineering: the hcome methodology. Int. J. Knowl. Inf. Syst. (KAIS) **10**, 109–131 (2006). http://www.icsd.aegean.gr/kotis/publications/HCOME-KAIS.pdf

174. Kotis, K., Vouros, G., Alonso, J.P.: Hcome: tool-supported methodology for collaboratively devising living ontologies. In: K. Kotis (ed.) Semantic Web and Databases, Workshop in VLDB'04 Conference. Toronto, Canada (2004). http://www.icsd.aegean.gr/kotis/publications/SWDB2004.pdf

175. Lattanzi, S., Sivakumar, D.: Affiliation networks. In: M. Mitzenmacher (ed.) Proceedings of the 41st Annual ACM Symposium on Theory of Computing, STOC 2009, pp. 427–434. ACM, New York, NY, USA (2009)

176. Lee, T., hoon Lee, I., Lee, S., goo Lee, S., Kim, D., Chun, J., Lee, H., Shim, J.: Building an operational product ontology system. Electron. Commerce Res. Appl. **5**(1), 16 – 28 (2006). https://doi.org/10.1016/j.elerap.2005.08.005, http://www.sciencedirect.com/science/article/pii/S1567422305000785. International Workshop on Data Engineering Issues in E-Commerce (DEEC 2005)

177. Lenat, D.B., Guha, R.V., Pittman, K., Pratt, D., Shepherd, M.: Cyc: Toward programs with common sense. Commun. ACM **33**(8), 30–49 (1990). http://dblp.uni-trier.de/db/journals/cacm/cacm33.html#LenatGPPS90

178. Leung, N.K.Y., Lau, S.K., Fan, J., Tsang, N.: An integration-oriented ontology development methodology to reuse existing ontologies in an ontology development process. In: Proceedings of the 13th International Conference on Information Integration and Web-based Applications and Services, iiWAS '11, pp. 174–181. ACM, New York, NY, USA (2011). https://doi.org/10.1145/2095536.2095567

179. Lewis, R.: Dereferencing http uris. Draft Tag Finding (May 31, 2007 (retrieved July 25, 2007). http://www. w3. org/2001/tag/doc/httpRange-14/2007-05-31/HttpRange-14.html (2007)

180. López, M.F., Gómez-Pérez, A., Sierra, J.P., Sierra, A.P.: Building a chemical ontology using methontology and the ontology design environment. IEEE Intell. Syst. Appl. **14**(1), 37–46 (1999)

181. Lozano-Tello, A., Gomez-Perez, A.: ONTOMETRIC: a method to choose the appropriate ontology. J. Database Manag. **15**(2), 1–18 (2004)

182. Madnick, S.E., Lupu, M.: Implementing the content interchange (coin) approach through use of semantic web tools. SSRN eLibrary (2008). https://doi.org/10.2139/ssrn.1269046

183. Madnick, S.E., Wernerfelt, B., Firat, A., Firat, A.: Information integration using contextual knowledge and ontology merging. Technical report (2003)

184. Maedche, A., Staab, S.: Measuring Similarity between Ontologies. In: Proceedings of the 13th International Conference on Knowledge Engineering and Knowledge Management: Ontologies and the Semantic Web (EKAW), pp. 251–263. Springer, London, UK (2002)

185. Maedche, A., Zacharias, V.: Clustering ontology-based metadata in the semantic web. In: Proceedings of the 6th European Conference on Principles of Data Mining and Knowledge Discovery (PKDD). Lecture Notes in Computer Science, vol. 2431, pp. 383–408. Springer, Berlin/Heidelberg, Helsinki, Finland (2002)

186. Manaf, N.A.A., Bechhofer, S., Stevens, R.: A survey of identifiers and labels in OWL ontologies. In: Sirin, E., Clark, K. (eds.) OWLED, CEUR Workshop Proceedings, vol. 614. CEUR-WS.org (2010)

187. Mariolis, P.: Interlocking directorates and control of corporations: the theory of bank control. Soc. Sci. Q. **56**(3), 425–439 (1975)

188. Martnez-Prieto, M., Arias Gallego, M., Fernndez, J.: Exchange and consumption of huge RDF Data. In: Simperl, E., Cimiano, P., Polleres, A., Corcho, O., Presutti, V. (eds.) The Semantic Web: Research and Applications. Lecture Notes in Computer Science, vol. 7295, pp. 437–452. Springer, Berlin, Heidelberg (2012). http://dx.doi.org/10.1007/978-3-642-30284-8_36

189. Masolo, C., Borgo, S., Gangemi, A., Guarino, N., Oltramari, A.: Wonderweb deliverable d18 ontology library (final). Technical report, IST Project 2001-33052 WonderWeb: Ontology Infrastructure for the Semantic Web (2003)

190. McBride, B.: Jena: a semantic web toolkit. IEEE Internet Comput. **6**(6), 55–59 (2002)

191. McCuinness, D.L.: Ontologies Come of Age. Spinning the semantic web: bringing the World Wide Web to its full potential p. 171 (2005)

192. McGuinness, D.L., van Harmelen, F.: OWL web ontology language overview. Technical report, W3C—World Wide Web Consortium (2004). http://www.w3.org/TR/owl-features/. Accessed 15 July 2012

193. Medicale, A.: Issues in the structuring and acquisition of an ontology for medical language understanding. Meth Inf. Med. **34**, 15–24 (1995)

194. Meditskos, G., Bassiliades, N.: Dlejena: A practical forward-chaining owl 2 r reasoner combining jena and pellet. Web Semant. **8**(1), 89–94 (2010). https://doi.org/10.1016/j.websem.2009.11.001

195. Mendes, O., Abran, A., Québec, H.K.M.: Software engineering ontology: a development methodology. In: Metrics News, vol. 9, 2004. Citeseer (2004)

196. Mika, P., Meij, E., Zaragoza, H.: Investigating the semantic gap through query log analysis. In: Bernstein, A., Karger, D.R., Heath, T., Feigenbaum, L., Maynard, D., Motta, E., Thirunarayan, K. (eds.) International Semantic Web Conference. Lecture Notes in Computer Science, vol. 5823, pp. 441–455. Springer (2009). http://dblp.uni-trier.de/db/conf/semweb/iswc2009.html#MikaMZ09

197. Mizoguchi, R., Ikeda, M.: Towards ontology engineering. Technical Report AI-TR-96-1, The Institute of Scientific and Industrial Research, Osaka University (1996)

198. Möller, K.: Lifecycle models of data-centric systems and domains. Semant. Web J. (2012). http://www.semantic-web-journal.net/sites/default/files/swj125_1.pdf

199. Neches, R., Fikes, R., Finin, T., Gruber, T., Patil, R., Senator, T., Swartout, W.: Enabling technology for knowledge sharing. AI Magazine **12**(3), 36–56 (1991)

200. Newman, M.: The mathematics of networks. New Palgrave Encyclopedia Econ. **2** (2008)

201. Newman, M.E.J.: Coauthorship networks and patterns of scientific collaboration. Proc. Natl. Acad. Sci. USA **101**(Suppl 1), 5200–5205 (2004)

202. Newman, M.E.J., Strogatz, S.H., Watts, D.J.: Random graphs with arbitrary degree distributions and their applications. Phys. Rev. E (Statistical, Nonlinear, and Soft Matter Physics) **64**(2), 026118 (2001)

203. Nikolov, A., Uren, V., Motta, E.: Data linking: capturing and utilising implicit schema-level relations. In: Proceedings of the Linked Data on the Web (LDOW 2010) at 19th International World Wide Web Conference (WWW 2010). CEUR Workshop Proceedings, vol. 628, pp. 1–11. CEUR-WS.org, Raleigh, USA (2010)

204. Noy, N.F., Klein, M.: Ontology evolution: not the same as schema evolution. Knowl. Inf. Syst. **6**(4), 428–440 (2004). http://www.cs.vu.nl/mcaklein/papers/NoyKlein.pdf

205. Noy, N.F., McGuinness, D.L.: Ontology development 101: a guide to creating your first ontology. Technical report, Stanford Knowledge Systems Laboratory and Stanford Medical Informatics (2001)

206. Noy, N.F., Shah, N.H., Whetzel, P.L., Dai, B., Dorf, M., Griffith, N., Jonquet, C., Rubin, D.L., Storey, M.A.D., Chute, C.G., Musen, M.A.: BioPortal: Ontologies and integrated data resources at the click of a mouse. Nucleic Acids Res. **37**(Web-Server-Issue), 170–173 (2009)

207. Oberle, D., Staab, S., Volz, R.: Three dimensions of knowledge representation in wonderweb. KI **19**(1), 31 (2005). http://dblp.uni-trier.de/db/journals/ki/ki19.html#OberleSV05

208. Obrst, L., Ceusters, W., Mani, I., Ray, S., Smith, B.: The evaluation of ontologies. In: Baker, C.J.O., Cheung, K.H. (eds.) Semantic Web, pp. 139–158. Springer, US (2007). http://dx.doi.org/10.1007/978-0-387-48438-9-8

209. Okamoto, K., Chen, W., Li, X.Y.: Ranking of closeness centrality for large-scale social networks. In: Preparata, F., Wu, X., Yin, J. (eds.) Frontiers in Algorithmics. Lecture Notes in Computer Science, vol. 5059, pp. 186–195. Springer, Berlin/Heidelberg (2008)

210. Oliveira, M.: Gama, J.A.: An overview of social network analysis. Wiley Interdisciplinary Reviews. Data Mining Knowl. Discov. **2**(2), 99–115 (2012). https://doi.org/10.1002/widm. 1048

211. Oltramari, A., Gangemi, A., Guarino, N., Masolo, C.: Restructuring wordnet's top-level: The ontoclean approach. LREC2002, Las Palmas, Spain **68** (2002)

212. OpenLink Software: Virtuoso Open-Source Edition (2009). http://virtuoso.openlinksw.com/ dataspace/dav/wiki/Main/

213. O'Reilly, T.: What is web 2.0: Design patterns and business models for the next generation of software. (2005). http://www.oreillynet.com/pub/a/oreilly/tim/news/2005/09/30/what-is-web-20.html

214. Oren, E., Möller, K., Scerri, S., Handschuh, S., Sintek, M.: What are semantic annotations? Technical report, DERI Galway (2006). http://www.siegfried-handschuh.net/pub/2006/ whatissemannot2006.pdf

215. Page, L., Brin, S., Motwani, R., Winograd, T.: The pagerank citation ranking: Bringing order to the web. Technical report 1999-66, Stanford InfoLab (1999)

216. Palma, R., Hartmann, J., Haase, P.: Omv-ontology metadata vocabulary for the semantic web. Technical report, Universidad Politecnica de Madrid, University of Karlsruhe, 2008. Version 2.4 (2008). http://omv.ontoware.org

217. Pant, G., Srinivasan, P.: Learning to crawl: comparing classification schemes. ACM Trans. Inf. Syst. (TOIS) **23**(4), 430–462 (2005)

218. Pant, G., Srinivasan, P., Menczer, F.: Crawling the web. Web. Dynamics **2004**, 153–178 (2004)

219. Patel-Schneider, P.F., Hayes, P., Horrocks, I.: OWL: web ontology language semantics and abstract syntax. W3C Recommendation (2004). http://www.w3.org/TR/2004/REC-owl-semantics-20040210/

220. Porzel, R., Malaka, R.: A task-based approach for ontology evaluation. In: Proceedings of the 16th European Conference on Artificial Intelligence, Valencia, Spain. Citeseer (2004)

221. Presutti, V., Gangemi, A., David, S., de Cea, G., Surez-Figueroa, M., Montiel-Ponsoda, E., Poveda, M.: Neon deliverable d2. 5.1. a library of ontology design patterns: reusable solutions for collaborative design of networked ontologies. NeOn Project (2008). http://www.neon-project.org

222. Radatz, J., Geraci, A., Katki, F.: Ieee standard glossary of software engineering terminology. IEEE Standard **610121990**, 121990 (1990)

223. Raimond, Y., Abdallah, S.A., Sandler, M.B., Giasson, F.: The Music Ontology. In: Proceedings of the 8th International Conference on Music Information Retrieval (ISMIR), pp. 417–422. Austrian Computer Society, Vienna, Austria (2007)

224. Rector, A.: Semantic interoperability deployment and research roadmap. Barriers, approaches and research priorities for integrating biomedical ontologies (WP 6, Task 6.1)IST-27328-SSA pp. 1–69 (2008). www.semantichealth.org/DELIVERABLES/SemanticHEALTH_D6_1.pdf

225. Rector, A.L., Drummond, N., Horridge, M., Rogers, J., Knublauch, H., Stevens, R., Wang, H., Wroe, C.: OWL Pizzas: practical experience of teaching OWL-DL: common errors and common patterns. In: Proceedings of 14th International Conference on Knowledge Engineering and Knowledge Management (EKAW). Lecture Notes in Computer Science, vol. 3257, pp. 63–81. Springer, Berlin/Heidelberg, Whittlebury Hall, UK (2004)

226. Richardson, M., Prakash, A., Brill, E.: Beyond pagerank: machine learning for static ranking. In: Proceedings of the 15th International Conference on World Wide Web, pp. 707–715. ACM (2006)

227. Sahay, R.N.: An ontological framework for interoperability of health level seven (hl7) applications:the ppepr methodology and system. Ph.D. thesis, Digital Enterprise Research Institute, National University of Ireland, Galway (2013). http://hdl.handle.net/10379/3034

228. Sandhaus, E.: Abstract: Semantic technology at the new york times: lessons learned and future directions. In: Patel-Schneider, P.F., Pan, Y., Hitzler, P., Mika, P., 0007, L.Z., Pan, J.Z.,

Horrocks, I., Glimm, B. (eds.) International Semantic Web Conference (2). Lecture Notes in Computer Science, vol. 6497, p. 355. Springer (2010). http://dblp.uni-trier.de/db/conf/semweb/iswc2010-2.html#Sandhaus10

229. Sauermann, L., Cyganiak, R.: Cool URIs for the Semantic Web. W3C Interest Group Note, W3C (2008). http://www.w3.org/TR/cooluris/

230. Schreiber, A.T., Terpstra, P.: Sisyphus-vt: a commonkads solution. Int. J. Hum.-Comput. Stud. **44**(3-4), 373–402 (1996). http://dblp.uni-trier.de/db/journals/ijmms/ijmms44.html#SchreiberT96

231. Schreiber, G., Wielinga, B., Jansweijer, W.: The kactus view on the 'o' word. In: Skuce, D. (ed.) The 1995 International Joint Conference on AI: Montreal, Quebec, Canada: 1995, August, 20-25, Workshop on Basic Ontological Issues in Knowledge Sharing, pp. 15.1–15.10 (1995)

232. Schwaber, K., et al.: Scrum development process. In: Proceedings of the Workshop on Business Object Design and Implementation at the 10th Annual Conference on Object-Oriented Programming Systems, Languages, and Applications (OOPSLA'95) (1995)

233. Shadbolt, N., Hall, W., Berners-Lee, T.: The semantic web revisited. IEEE Intell. Syst. **21**(3), 96–101 (2006). https://doi.org/10.1109/MIS.2006.62

234. Simmons, E.: The usage model: a structure for richly describing product usage during design and development. In: Proceedings of 13th IEEE International Conference on Requirements Engineering, pp. 403–407. IEEE, Paris, France (2005)

235. Simperl, E.: Reusing ontologies on the Semantic Web: a feasibility study. Data Knowl. Eng. **68**(10), 905–925 (2009)

236. Sirin, E., Parsia, B.: Pellet: An owl dl reasoner. In: Description Logics (2004). http://sunsite.informatik.rwth-aachen.de/Publications/CEUR-WS//Vol-104/30Sirin-Parsia.pdf

237. Sleeman, J., Finin, T.: Learning Co-reference Relations for FOAF Instances. In: Proceedings of the International Semantic Web Conference (ISWC) Posters & Demonstrations Track. CEUR Workshop Proceedings, vol. 658. CEUR-WS.org, Shanghai, China (2010)

238. Smith, B., Ashburner, M., Rosse, C., Bard, J., Bug, W., Ceusters, W., Goldberg, L., Eilbeck, K., Ireland, A., Mungall, C., et al.: The obo foundry: coordinated evolution of ontologies to support biomedical data integration. Nat. Biotechnol. **25**(11), 1251–1255 (2007)

239. Sommerville, I.: Software Engineering, 6th edn. Pearson, Studium (2001)

240. Staab, S., Studer, R.: Handbook on Ontologies, 2nd edn. Springer, Berlin (2009)

241. Staab, S., Studer, R., Schnurr, H.P., Sure, Y.: Knowledge processes and ontologies. IEEE Intell. Syst. **16**(1), 26–34 (2001). http://dblp.uni-trier.de/db/journals/expert/expert16.html#StaabSSS01

242. Steve, G., Gangemi, A., Pisanelli, D.M.: Integrating medical terminologies with onions methodology. In: Information Modelling and Knowledge Bases VIII. IOS Press (1997)

243. Stojanovic, L.: Methods and tools for ontology evolution. Ph.D. thesis, Karlsruhe Institute of Technology (2004). http://digbib.ubka.uni-karlsruhe.de/volltexte/1000003270

244. Stojanovic, L., Maedche, A., Motik, B., Stojanovic, N.: User-driven ontology evolution management. Knowledge Engineering and Knowledge Management: Ontologies and the Semantic Web. Lecture Notes in Computer Science, vol. 2473, pp. 133–140. Springer, Berlin/Heidelberg, Siguenza, Spain (2002)

245. Stroetman, V., Kalra, D., Lewalle, P., Rector, A., Rodrigues, J., Stroetman, K., Surjan, G., Ustun, B., Virtanen, M., Zanstra, P.: Semantic interoperability for better health and safer healthcare [34 pages] (2009)

246. Stuckenschmidt, H.: Modularization of ontologies. WonderWeb Deliverable D21 (2003). http://wonderweb.semanticweb.org/deliverables/D21.shtml

247. Studer, R., Benjamins, V.R., Fensel, D.: Knowledge engineering: principles and methods. Data Knowl. Eng. **25**(1-2), 161–197 (1998). http://dx.doi.org/10.1016/S0169-023X(97)00056-6

248. Suárez-Figueroa, M.C., Gómez-Pérez, A., Fernández-López, M.: The NeOn Methodology for Ontology Engineering. In: Suárez-Figueroa, M.C., Gómez-Pérez, A., Motta, E., Gangemi, A. (eds.) Ontology Engineering in a Networked World, pp. 9–34. Springer, Berlin, Heidelberg (2012). http://dx.doi.org/10.1007/978 3 642-24794-12

249. Sure, Y.: On-to-knowledge: Ontology based knowledge management tools and their application. Kunstliche Intelligenz pp. 35–37 (2002)
250. Sure, Y., Akkermans, H., Broekstra, J., Davies, J., Ding, Y., Duke, A., Engels, R., Fensel, D., Horrocks, I., Iosif, V., et al.: On-to-knowledge: semantic web enabled knowledge management. Web Intell. **35**, 277–300 (2003)
251. Sure, Y., Erdmann, M., Angele, J., Staab, S., Studer, R., Wenke, D.: Ontoedit: Collaborative ontology development for the semantic web. In: Horrocks, I., Hendler, J.A. (eds.) International Semantic Web Conference. Lecture Notes in Computer Science, vol. 2342, pp. 221–235. Springer (2002). http://dblp.uni-trier.de/db/conf/semweb/semweb2002.html#SureEASSW02
252. Sure, Y., Erdmann, M., Angele, J., Staab, S., Studer, R., Wenke, D.: OntoEdit: Collaborative ontology development for the semantic web. In: Horrocks, I., Hendler, J.A. (eds.) International Semantic Web Conference. Lecture Notes in Computer Science, vol. 2342, pp. 221–235. Springer (2002)
253. Sure, Y., Staab, S., Studer, R.: On-To-Knowledge Methodology (OTKM). In: Staab, S., Studer, R. (eds.) Handbook on Ontologies: International Handbook on Information Systems, pp. 117–132. Springer (2004)
254. Swartout, B., Patil, R., Knight, K., Russ, T.: Toward distributed use of large-scale ontologies. In: Ontological Engineering, AAAI-97 Spring Symposium Series, pp. 138–148 (1997)
255. T.B. Lee, J.H., Lassila., O.: The semantic web. In: Scientific America, p. 284 (2001)
256. Tao, C., Embley, D.W., Liddle, S.W.: Focih: Form-based ontology creation and information harvesting. In: Laender, A.H.F., Castano, S., Dayal, U., Casati, F., de Oliveira, J.P.M. (eds.) ER. Lecture Notes in Computer Science, vol. 5829, pp. 346–359. Springer (2009)
257. Tartir, S., Arpinar, I.B.: Ontology evaluation and ranking using ontoqa. In: International Conference on Semantic Computing **0**, 185–192 (2007). https://doi.org/10.1109/ICSC.2007.19
258. Tartir, S., Arpinar, I.B., Moore, M., Sheth, A.P., Aleman-Meza, B.: OntoQA: metric-based ontology quality analysis. In: Proceedings of IEEE Workshop on Knowledge Acquisition from Distributed, Autonomous, Semantically Heterogeneous Data and Knowledge Sources, vol. 9, pp. 45–53. IEEE Computer Society Press, Houston, Texas (2005)
259. Thelwall, M., Stuart, D.: Web crawling ethics revisited: Cost, privacy, and denial of service. J. Am. Soc. Inf. Sci. Technol. **57**(13), 1771–1779 (2006). http://dx.doi.org/10.1002/asi.20388
260. Tong, A., Drees, B., Nardelli, G., Bader, G., Brannetti, B., Castagnoli, L., Evangelista, M., Ferracuti, S., Nelson, B., Paoluzi, S., et al.: A combined experimental and computational strategy to define protein interaction networks for peptide recognition modules. Sci. Signal. **295**(5553), 321 (2002)
261. TopQuadrant: TopQuadrant I Products I TopBraid Composer (2011). http://www.topquadrant.com/products/TB_Composer.html
262. Tran, T., Haase, P., Lewen, H., Óscar Muñoz García, Gómez-Pérez, A., Studer, R.: Lifecycle-support in architectures for ontology-based information systems. pp. 508–522 (2008). https://doi.org/10.1007/978-3-540-76298-07
263. Tsarkov, D., Horrocks, I.: Fact++ description logic reasoner: System description. Automated Reasoning pp. 292–297 (2006)
264. Tudorache, T., Falconer, S., Nyulas, C., Noy, N.F., Musen, M.A.: Will semantic web technologies work for the development of icd-11? In: The Semantic Web–ISWC 2010, pp. 257–272. Springer (2010)
265. Tummarello, G., Cyganiak, R., Catasta, M., Danielczyk, S., Delbru, R., Decker, S.: Sig.ma: live views on the web of data. Web Semant.: Sci. Serv. Agents. World Wide Web **8**(4), 355–364 (2010)
266. Tummarello, G., Oren, E., Delbru, R.: Sindice.com: Weaving the open linked data. In: Proceedings of the 6th International Semantic Web Conference and 2nd Asian Semantic Web Conference (ISWC/ASWC), Busan, South Korea. LNCS, vol. 4825, pp. 547–560. Springer, Berlin, Heidelberg (2007)

267. Tutoky, G.: Collaboration Social Networks Information Sciences and Technologies. In: Bulletin of the ACM Slovakia - ISSN 1338-1237. ACM (2011)
268. Uschold, M.: Building ontologies: towards a unified methodology. In: Proceedings of Expert Systems '96, the 16th Annual Conference of the British Computer Society Specialist Group on Expert Systems. Cambridge, UK (1996)
269. Uschold, M., Bateman, J., Davis, M., Sowa, J.: Ontology summit 2011 communique : Making the case for ontology (2011). http://ontolog.cim3.net/file/work/OntologySummit2011/
270. Uschold, M., Gruninger, M.: Ontologies: principles, methods and applications. Knowl. Eng. Rev. 11(02), 93–136 (1996). https://doi.org/10.1017/S0269888900007797
271. Uschold, M., Jasper, R.: A framework for understanding and classifying ontology applications. In: Proceedings of the IJCAI-99 Workshop on Ontologies and Problem-Solving Methods (KRR5), Stockholm, Sweden, 2 Aug 1999
272. Uschold, M., King, M.: Towards a methodology for building ontologies. In: Workshop on Basic Ontological Issues in Knowledge Sharing, held in conjunction with IJCAI-95. Montreal, Canada (1995)
273. Valente, A., Russ, T., MacGregor, R., Swartout, W.: Building and (re)using an ontology of air campaign planning. IEEE Intell. Syst. Appl. 14(1), 27–36 (1999). https://doi.org/10.1109/5254.747903
274. Van Damme, C., Hepp, M., Siorpaes, K.: Folksontology: An integrated approach for turning folksonomies into ontologies. In: Bridging the Gap between Semantic Web and Web2.0 (2007). http://www.kde.cs.uni-kassel.de/ws/eswc2007/proc/FolksOntology.pdf
275. Vega, J.C.A., Corcho, O., Fernández-López, M., Gómez-Pérez, A.: WebODE: A scalable workbench for ontological engineering. In: K-CAP, pp. 6–13. ACM (2001). http://dblp.uni-trier.de/db/conf/kcap/kcap2001.html#VegaCFG01
276. Völker, J., Vrandecic, D., Sure, Y.: Automatic Evaluation of Ontologies (AEON). In: Gil, Y., Motta, E., Benjamins, V.R., Musen, M.A. (eds.) Proceedings of the 4th International Semantic Web Conference (ISWC'05). LNCS, vol. 3729, pp. 716–731. Springer, Galway, Ireland (2005)
277. Volz, R.: Ontoserverinfrastructure for the semantic web. SWWS01 p. 96 (2001)
278. Volz, R., Oberle, D., Motik, B., Staab, S., Studer, R.: Kaon server architecture. Technical report D5, WonderWeb project deliverable (2002). Also published as AIFB technical report
279. Vossen, G., Hagemann, S.: Unleashing Web 2.0: From Concepts to Creativity. Morgan Kaufmann (2007). http://www.amazon.com/exec/obidos/redirect?tag=citeulike07-20&path=ASIN/0123740347
280. Vrandecic, D.: Ontology evaluation. Ph.D. thesis, Karlsruhe Institute of Technology (2010). http://www.aifb.kit.edu/images/b/b5/OntologyEvaluation.pdf
281. Vrandečić, D., Gangemi, A.: Unit Tests for Ontologies. In: Meersman, R., Tari, Z., Herrero, P. (eds.) On the Move to Meaningful Internet Systems 2006: OTM 2006 Workshops. Lecture Notes in Computer Science, vol. 4278, pp. 1012–1020. Springer, Berlin, Heidelberg (2006). http://dx.doi.org/10.1007/11915072_2
282. Vrandecic, D., Pinto, S., Tempich, C., Sure, Y.: The DILIGENT knowledge processes. J. Knowl. Manag. 9(5), 85–96 (2005)
283. Vrandečić, D., Sure, Y.: How to design better ontology metrics. The Semantic Web: Research and Applications pp. 311–325 (2007)
284. Wasserman, S., Faust, K.: Social Network Analysis: Methods and Applications, vol. 8, 1st edn. In: Structural analysis in the Social Sciences. Cambridge University Press (1994)
285. Weisstein, E.: Normal distribution. In From MathWorld, A Wolfram Web Resource (2005). http://mathworld.wolfram.com/NormalDistribution.html
286. Weisstein, E., Polynomials, M.: Mathworld–a wolfram web resource. From MathWorld, A Wolfram Web Resource (2004). http://mathworld.wolfram.com/BellNumber.html
287. Wielinga, B., Akkermans, J.M., Schreiber, A.T.: A formal analysis of parametric design problem solving. In: Proceedings of the 9th Banff Knowledge Acquisition Workshop (KAW-95), pp. 37–41 (1995)
288. Wongthongtham, P., Chang, E., Dillon, T.S., Sommerville, I.: Ontology-based multi-site software development methodology and tools. J. Syst. Architect. 52(11), 640–653 (2006)

289. Zablith, F.: Ontology evolution: a practical approach. In: Proceedings of Workshop on Matching and Meaning at Artificial Intelligence and Simulation of Behaviour (AISB), Edinburgh, UK. AISB Workshops, vol. 1 (2009)
290. Zablith, F.: Harvesting online ontologies for ontology evolution. Ph.D. thesis, The Knowledge Media Institute (KMi), The Open University, Milton Keynes, The United Kindgom (2011). http://fouad.zablith.org
291. Zachary, W.: An information flow model for conflict and fission in small groups. J. Anthropol. Res. **33**, 452–473 (1977)
292. Zhang, H.: The scale-free nature of semantic web ontology. In: Proceeding of the 17th International Conference on World Wide Web, pp. 1047–1048. ACM (2008)
293. Zhang, X., Li, H., Qu, Y.: Finding Important Vocabulary Within Ontology. In: Mizoguchi, R., Shi, Z., Giunchiglia, F. (eds.) ASWC. Lecture Notes in Computer Science, vol. 4185, pp. 106–112. Springer (2006)
294. Zhou, D., Huang, J., Schälkopf, B.: Learning with hypergraphs: clustering, classification, and embedding. In: Schälkopf, B., Platt, J., Hoffman, T. (eds.) NIPS, pp. 1601–1608. MIT Press (2006)
295. Zimmermann, A.: Ontology recommendation for the data publishers. In: Proceedings of 1st Workshop on Ontology Repositories. CEUR Workshop Proceedings, vol. 596. CEUR-WS.org, Aachen, Germany (2010)
296. Zweigenbaum, P.: Menelas: an access system for medical records using natural language. Comput. Methods Progr. Biomed. **45**(1–2), 117–120 (1994)
297. Zweigenbaum, P., Jacquemart, P., Grabar, N., Habert, B.: Building a text corpus for representing the variety of medical language. Stud. Health Technol. Inf. **1**, 290–294 (2001)

Printed in the United States
By Bookmasters